✹ Smithsonian

Space

a visual encyclopedia

Second Edition

DK Delhi
Senior Editor Virien Chopra
Senior Art Editor Vikas Chauhan
Project Art Editor Heena Sharma
Editor Kathakali Banerjee
Managing Editor Kingshuk Ghoshal
Managing Art Editor Govind Mittal
DTP Designers Bimlesh Tiwary, Vikram Singh
Picture Researcher Aditya Katyal
Jacket Designer Tanya Mehrotra
Picture Research Manager Taiyaba Khatoon
Pre-Production Manager Balwant Singh
Production Manager Pankaj Sharma

DK London
Senior Art Editor Louise Dick
Project Editor Francesco Piscitelli
US Editor Kayla Dugger
US Executive Editor Lori Cates Hand
Managing Editor Lisa Gillespie
Managing Art Editor Owen Peyton Jones
Senior Production Editor Andy Hilliard
Senior Production Controller Meskerem Berhane
Jacket Design Development Manager Sophia MTT
Publisher Andrew Macintyre
Associate Publishing Director Liz Wheeler
Art Director Karen Self
Publishing Director Jonathan Metcalf
Consultants Giles Sparrow, Ian Ridpath

First Edition
Project Editor Wendy Horobin
Project Designer Pamela Shiels
Editors Fleur Star, Holly Beaumont, Lee Wilson, Susan Malyan
Designers Rachael Grady, Lauren Rosier, Gemma Fletcher, Karen Hood,
Clare Marshall, Mary Sandberg, Sadie Thomas
US Editor Margaret Parrish
Indexer Chris Bernstein
Picture Researchers Ria Jones, Harriet Mills, Rebecca Sodergren
Production Editor Sean Daly
Jacket Designer Natalie Godwin
Jacket Editor Matilda Gollon
Publishing Manager Bridget Giles
Art Director Martin Wilson
Consultant Peter Bond

Packaging services supplied by **Bookwork**

This American Edition, 2020
First American Edition, 2010
Published in the United States by DK Publishing
1450 Broadway, Suite 801, New York, NY 10018

Copyright © 2010, 2020 Dorling Kindersley Limited
DK, a Division of Penguin Random House LLC
20 21 22 23 24 10 9 8 7 6 5 4 3 2 1
001–317626–Aug/2020

A catalog record for this book is available from the Library of Congress.
ISBN 978-1-4654-9425-2

DK books are available at special discounts when purchased in bulk
for sales promotions, premiums, fund-raising, or educational use.
For details, contact: DK Publishing Special Markets,
1450 Broadway, Suite 801, New York, NY 10018
SpecialSales@dk.com

Printed and bound in China

For the curious

www.dk.com

Established in 1846, the Smithsonian—the world's largest museum and research
complex—includes 19 museums and galleries and the National Zoological Park. The
total number of artifacts, works of art, and specimens in the Smithsonian's collection is
estimated at 156 million. The Smithsonian is a renowned research center, dedicated to
public education, national service, and scholarship in the arts, sciences, and history.

Contents

Introduction

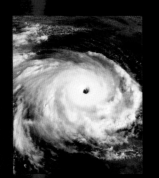

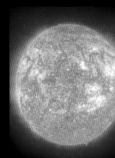

More than 500 people have left Earth behind to explore the wonders of outer space. Now you, too, can voyage through space and time and enjoy an experience that is out of this world.

As you turn the pages of this lavishly illustrated encyclopedia, you will learn about how rockets and telescopes work, discover what it is like to work and live in space, and unravel the mysteries of the final frontier. You'll travel from our small, blue planet to strange worlds with poisonous atmospheres, hidden oceans, and huge volcanoes. Then head out into the Milky Way to discover the multicolored clouds, stars, and galaxies that lie scattered throughout the universe.

Packed with beautiful images from the world's most powerful telescopes and full of amazing facts, this encyclopedia is invaluable as a reference book for researching projects or perfect for just dipping into.

For anyone who has ever stared up at the night sky and wondered what the universe is really like, this book is an essential read.

Peter Bond

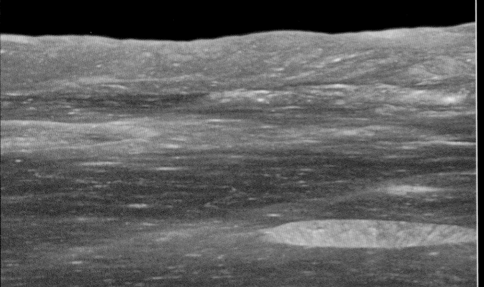

When you see this symbol in the book, turn to the pages listed to find out more about a subject.

▲ GENERAL ARTICLES *focus on particular topics of interest (* *p72–73). Many have fact boxes, timelines that chronicle key stages in development, and picture features.*

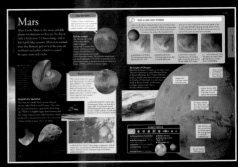

▲ DETAILED PROFILES *accompany features about our solar system (* *p128–129). These are packed with facts and figures about the structure, composition, and features of each planet.*

▲ FACT FILES *take an in-depth look at one topic, such as telescopes (* *p18–19). They detail all you need to know about the subject.*

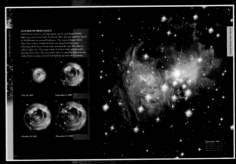

▲ PHOTO SPREADS *capture items of special interest within each chapter, such as exploding stars (* *p216–217).*

5

OBSERVING
THE UNIVERSE

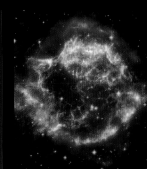

We live on one tiny planet in the vast universe. Finding out what else is "out there" has been one of our biggest challenges, and it started with people simply gazing at the sky.

What is *space?*

We live on a small blue planet called Earth. It has a surface of liquid water and rock and is surrounded by a blanket of air called the atmosphere. Space begins at the top of the atmosphere. It is an unimaginably vast, silent, and largely empty place, but it has many amazing properties.

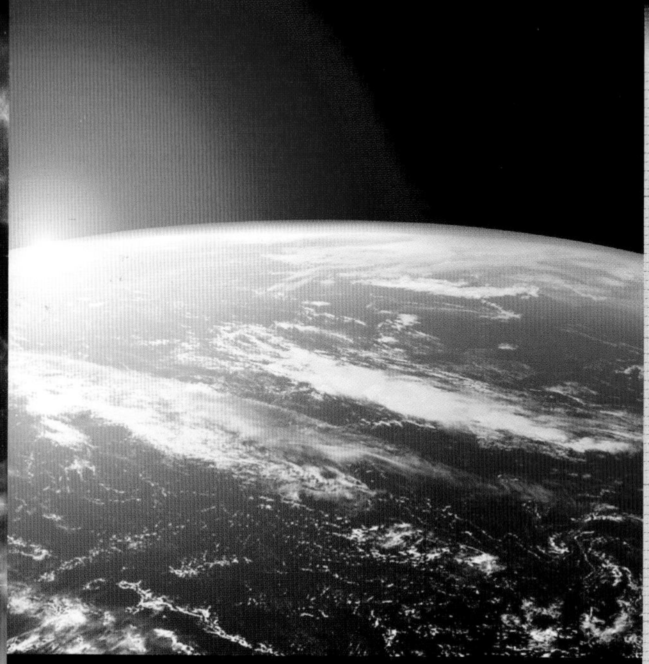

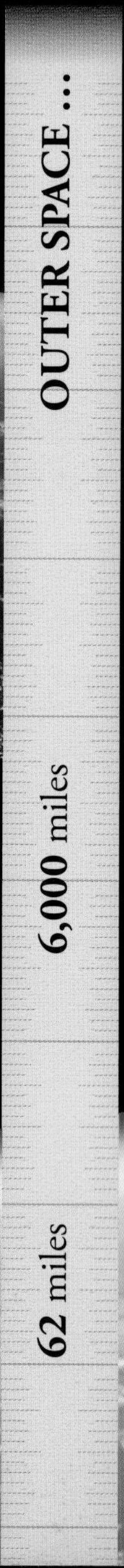

OUTER SPACE ...

◄ OUTER SPACE *Even in places far from stars and planets, space contains some scattered particles and dust.*

THE EDGE OF SPACE

Earth's atmosphere does not end suddenly—it gets gradually thinner and thinner as you travel up from the ground. Most experts agree that outer space starts at a height of 62 miles (100 km). Yet even above this height, there is a layer of very thin air called the exosphere. Hydrogen and other light gases are slowly escaping into space from this outermost part of Earth's atmosphere.

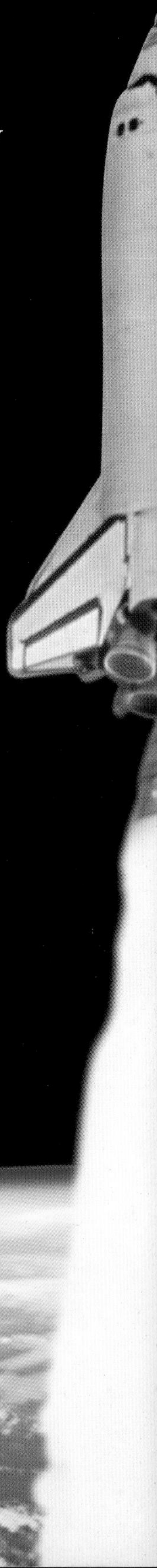

6,000 miles

◄ EXOSPHERE *This top layer of the atmosphere extends up to 6,000 miles (10,000 km) above Earth.*

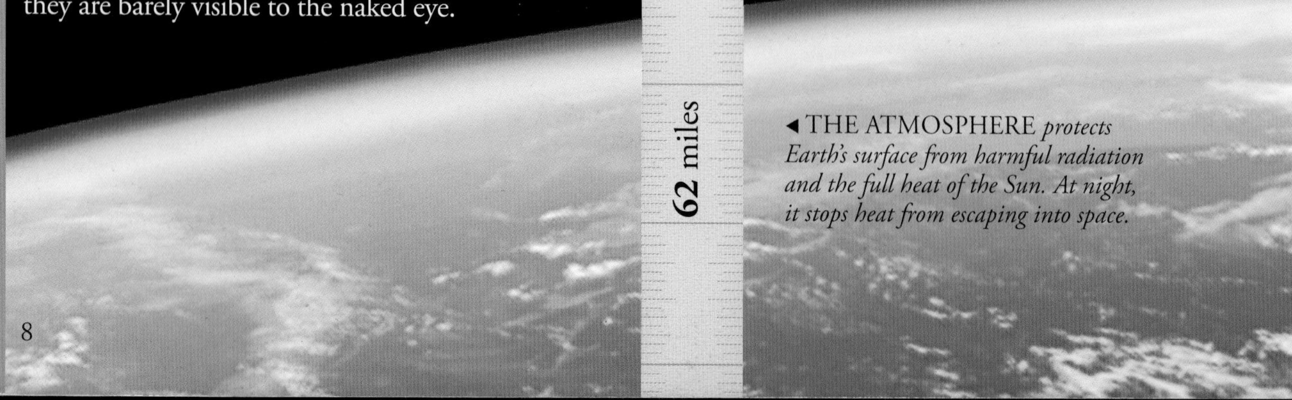

62 miles

Blacker than black

Planets like Earth shine because they reflect light from the Sun. Stars shine because they produce huge amounts of energy by burning fuel. In photos taken from space, our planet is surrounded by blackness. Most of space looks dark or black because the universe is very vast, and light from distant objects is traveling from so far away that they are barely visible to the naked eye.

◄ THE ATMOSPHERE *protects Earth's surface from harmful radiation and the full heat of the Sun. At night, it stops heat from escaping into space.*

IN A VACUUM

A place without any air or gas is called a vacuum. On Earth, air transfers heat from one place to another. As there is very little gas to distribute heat in the emptiness of space, it is nearly a vacuum. In space, the sunlit side of a spacecraft gets very hot, while the other side is in darkness and gets very cold. Spacecraft have to be tested in a thermal vacuum chamber before they are launched to make sure that they can survive these extreme space temperatures.

Sun

☐ WATCH THIS SPACE

Anything that travels through space at a steady speed is weightless. This is why things inside a spacecraft float and why astronauts are able to lift huge satellites using just their hands. The weightlessness disappears if the spacecraft either slows down or speeds up.

Hot

▶ BARBECUE ROLL *This is a slow, rotating movement used to stop any part of a spaceship from getting too hot or too cold.*

Cold

▲ ESCAPING GRAVITY
Spacecraft like the US space shuttle use up nearly all of their rocket fuel just to overcome gravity and reach outer space.

Getting off the ground

It is difficult to get into space because Earth's gravity holds everything down. To overcome gravity and go into orbit, a rocket has to reach a speed of 17,500 mph (28,000 kph), which requires a lot of fuel to provide energy. To reach the Moon and planets, spacecraft have to travel at an even higher speed—25,200 mph (40,000 kph), known as escape velocity.

Our place in space

Planet Earth is our home, and to us it seems like a very big place. Flying to the other side of the world takes an entire day, and sailing around the world takes many weeks. Yet in the vastness of the Universe, Earth is just a tiny dot. In fact, an alien flying through our galaxy would probably not even notice our little planet.

EARTH AND MOON
Earth's nearest neighbor is the Moon, our planet's only natural satellite. The Moon is a lot smaller than Earth. Its diameter is only about one-quarter the diameter of Earth, and 50 Moons would fit inside Earth. Although it looks quite close, the Moon is actually about 240,000 miles (384,000 km) away. It would take a crewed spacecraft 3 days to travel from Earth to the Moon.

THE SOLAR SYSTEM
Earth is just one of many objects that orbit the star we call the Sun. The Sun's "family" consists of eight planets, many dwarf planets, hundreds of moons, millions of comets and asteroids, and lots of gas and dust. All these things together are called the solar system. The four small planets closest to the Sun are made of rock, while the four outer planets are much larger and made mostly of gas, liquid, and ice. The solar system is big—the Voyager spacecraft took 12 years to reach Neptune, the outermost planet.

THE LOCAL GROUP

The Milky Way is one of the largest galaxies in a cluster of over 50 galaxies known as the Local Group. Most of these galaxies are loose balls or clouds of stars and are much smaller than the Milky Way. The two closest galaxies to the Milky Way are called the Large and the Small Magellanic Clouds. They lie about 200,000 light-years away and are easily visible with the naked eye from Earth's southern hemisphere. The biggest galaxy in the Local Group is the Andromeda Galaxy—a great spiral galaxy, much like the Milky Way. It lies about 2.5 million light-years away, in the constellation of Andromeda.

THE UNIVERSE

The Universe is everything that exists—all the stars, planets, galaxies, and space between them. There are millions of galaxy clusters in the Universe; in fact, wherever we look with telescopes, the sky is full of galaxies. Scientists estimate that there must be about 10 thousand billion billion stars in the Universe we can see—more than the number of grains of sand on all the beaches on Earth.

THE MILKY WAY

Our solar system is located in a large spiral-shaped galaxy called the Milky Way. The Sun is just one of at least 100 billion stars in this galaxy and lies about 26,000 light-years from the center. The Milky Way is vast—it measures more than 100,000 light-years across. A spaceship traveling at the speed of light (186,000 miles or 300,000 km per second) would take at least 100,000 years to fly from one side of the galaxy to the other. Stars in our part of the Milky Way are a long way apart—the nearest star to our Sun is more than 4 light-years away.

FAST FACTS

- It would take a modern jet fighter more than a million years to reach the nearest star.
- A light-year is the distance that light travels in 1 year. It is about 6 trillion, or 6 million million miles (9.5 trillion km).
- How big is the Universe? No one knows, because we cannot see the edge of it— if there is one. All we do know is that the visible Universe is at least 93 billion light-years across.
- The Universe has no center.

11

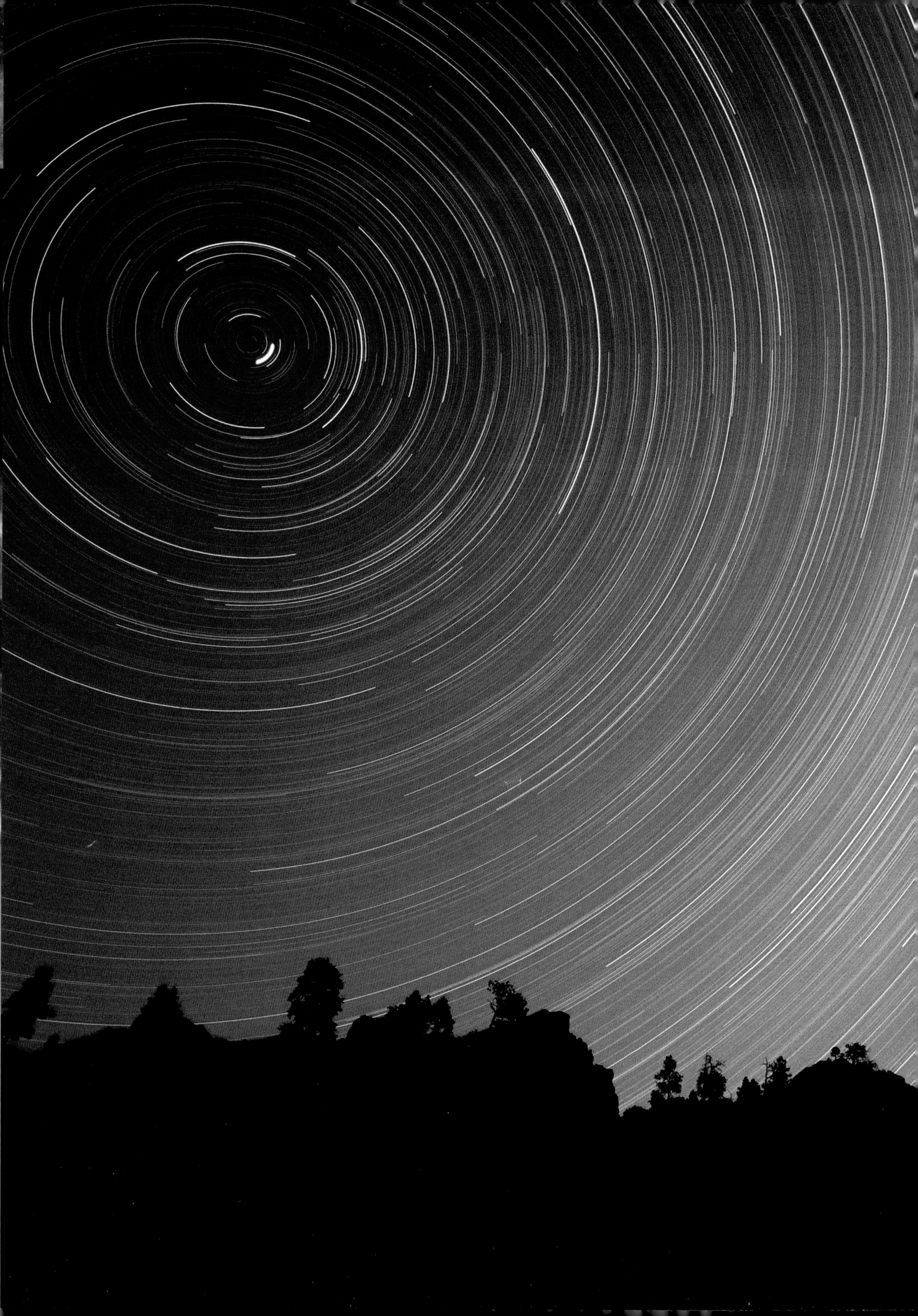

A CIRCLE OF STARS

This time-exposure photograph was taken in late summer in British Columbia, Canada. The circular lines of light are the trails of northern polar stars. However, the stars are not moving—the trails appear because the camera gradually moves as Earth rotates on its axis.

Early ideas

Compared to everything else around us, Earth seems incredibly large. Ancient peoples believed it was the biggest and most important place in the Universe and that everything revolved around it. These ideas only began to change very slowly after the introduction of the telescope in the early 1600s.

EARTH-CENTERED UNIVERSE

Ancient peoples watched the Sun, Moon, and stars very carefully. They saw that all of them traveled from east to west across the sky. Clearly, they were all going around a stationary Earth. For several thousand years, almost everyone believed that Earth was at the center of the Universe. The main problem with this idea was that it did not explain the movements of some of the planets—sometimes Mars or Jupiter appeared to stand still or even move backward.

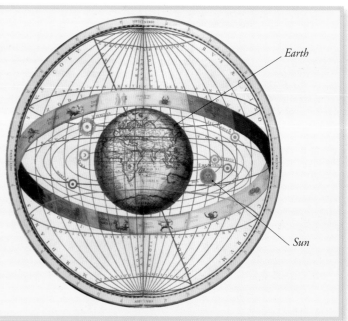

Earth

Sun

FLAT EARTH OR ROUND EARTH?

Stand on the seashore and look at the horizon. It seems to be flat. For a long time, people thought that Earth was flat and that, if you went too far, you would fall off the edge. However, it was gradually realized that Earth was round, like a giant ball. Nature provided several clues:

■ The shadow that Earth casts on the Moon during a lunar eclipse is curved, not straight.
■ A sailor traveling due north or south sees stars appear and disappear over the horizon. On a flat Earth, he would always see the same stars.
■ A ship sailing over the horizon should simply get smaller and smaller if Earth is flat. In fact, the hull disappears first, and the top of the sails last.

▼ LAND AHOY! *As the boat gets closer to the island, the sailor sees the tops of the mountains first. Then, as the boat moves over the curve, lower land comes into view.*

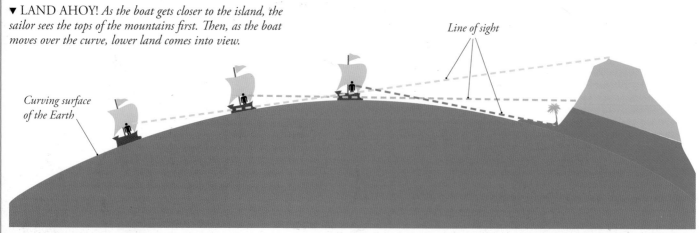

Line of sight

Curving surface of the Earth

ORBITS

The ancient Greeks taught that the circle was the perfect shape, so it seemed logical to believe that all of the planets traveled in circles. Unfortunately, measurements showed that this did not fit their movements across the sky. One way around this was to add small circles to the larger circles, but even this did not work. The mystery was solved in 1609 when a German mathematician, Johannes Kepler, realized that the planets move along elliptical (oval) orbits.

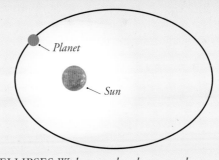

Planet

Sun

▶ *Johannes Kepler*

▲ ELLIPSES *We know today that most planets orbit the Sun in a path that is not quite circular. Pluto's orbit is a very stretched circle known as an ellipse.*

📷 WHAT A STAR!

Polish astronomer Nicolaus Copernicus (1473–1543) argued that the Sun, not Earth, is at the center of the solar system. His ideas were extremely unpopular.

◀ *A lunar eclipse*

▶ HE SAW IT COMING
Hipparchus developed a way of predicting solar and lunar eclipses using mathematical calculations.

Astounding astronomer

One of the greatest early Greek astronomers was Hipparchus of Nicaea (190–120 BCE). He discovered many things, including that Earth rotated on a tilted axis, which caused the seasons. He worked out the distance from Earth to the Moon by comparing views of a partial and total solar eclipse. He found that the Moon had an elliptical orbit and that its speed varied. He also cataloged the naked-eye stars in order of brightness. Another Greek, Ptolemy, later arranged the stars into 48 constellations in the 2nd century CE.

THE CALENDAR

Different ancient civilizations built tools to study the positions of celestial objects in the sky, such as the Sun and the stars. They used the movement of the Sun as their calendar and built monuments and temples that reflected the calendar. The Toltecs of Central America built the Pyramid of Kukulcan with 365 steps—one for each day of the year.

▲ PYRAMID OF KUKULCAN *This temple is built in honor of Kukulcan, the serpent god. When the Sun is in the right place, it creates a shadow in the shape of a serpent.*

Telescopes

Telescopes are instruments for looking at things that are far away. We've learned a lot about space by looking through telescopes. Optical telescopes can capture light from the deepest parts of space but are limited by the size of their mirrors and lenses.

▶ THE YERKES OBSERVATORY *was funded by business tycoon Charles T. Yerkes, who had developed Chicago's transportation system.*

REFRACTING TELESCOPE

The first telescopes were refracting telescopes, which used lenses to bend and focus light. The biggest refracting telescope is at Yerkes Observatory in Wisconsin. Built in 1897, it can be used to look at stars and track their movements through space.

▼ THE YERKES TELESCOPE
Built in 1897, the Yerkes telescope has a lens diameter of 40 in (100 cm) and weighs 6 tons (5.5 tonnes)—as much as an adult African elephant.

Objective lens

Eyepiece

Lenses magnify the image

Refracting telescopes

A refracting telescope uses a convex (outward-curving) glass lens to collect and focus incoming light. An eyepiece is used to magnify the image. One problem with using lenses is that they are heavy. If they are too big, they will start to sag, distorting the image. This limits the size and power of the refracting telescope.

Eyepiece
Small mirror

Objective mirror

Reflecting telescopes

A concave (inward-curving) mirror focuses light toward a smaller mirror. This sends the beam of light to an eyepiece, which magnifies the image. Because mirrors are lighter than lenses, reflecting telescopes can be much bigger and more powerful than refracting telescopes.

EVEN BIGGER TELESCOPES

Although reflecting telescopes can be built much bigger than refracting telescopes, they, too, will have problems if the mirror is more than 27.5 ft (8 m) across. Astronomers solve this problem by using a number of smaller mirrors that can be fitted together to make one big mirror. Each mirror section is controlled by a computer that can adjust its position by less than the width of a human hair.

Did you know…

Not all telescopes use glass mirrors—some use liquid metal instead. A shallow bowl of mercury or silver is spun at high speed until it forms a thin reflective surface. Liquid mirrors can only be used to look straight up. If they are tilted, the liquid will fall out!

🔍 TAKE A LOOK: EARLY TELESCOPES

The first telescopes were made by Dutch spectacle-maker Hans Lippershey in 1608. These were simple refracting telescopes made from a pair of glass lenses set into a tube. When the Italian astronomer Galileo Galilei heard about Lippershey's invention, he quickly set about building an improved telescope with a greater magnification.

▲ HANS LIPPERSHEY *is said to have come up with his invention while watching two young boys playing with lenses.*

▶ GALILEO'S DRAWINGS *By 1610, Galileo had developed a much more powerful telescope. He used this to study the surface of the Moon, the phases of Venus, and the moons of Jupiter.*

◀ NEWTON'S TELESCOPE *Isaac Newton made the first working reflector telescope in 1668.*

Giant *telescopes*

The Hale telescope caused quite a stir when it was completed in 1948. Equipped with a 16 ft (5 m) mirror, it was the largest and most powerful telescope ever built. As technology has improved, telescopes have been built with mirrors around 33 ft (10 m) across. Even larger telescopes are now planned, with mirrors of 98 ft (30 m) or more.

TELL ME MORE ...

To get the best images, telescopes are placed at high altitude so they are above the clouds and most of the atmosphere. Remote mountains are ideal, as there is little light interference from nearby towns. Mauna Kea, an extinct volcano in Hawaii, is home to many telescopes.

Keck Telescopes

- **Size of primary mirrors** 33 ft (10 m) each
- **Location** Mauna Kea, Hawaii
- **Altitude** 13,600 ft (4,145 m)

Until 2009, the twin Keck telescopes were the world's largest optical telescopes. The Keck II telescope overcomes the distorting effects of the atmosphere by using a mirror that changes shape 2,000 times per second.

Gemini Telescopes

- **Size of primary mirrors** 26 ft (8 m) each
- **Location** North: Mauna Kea, Hawaii; South: Cerro Pacho, Chile
- **Altitude** North: 13,822 ft (4,213 m); South: 8,930 ft (2,722 m)

The twin Gemini telescopes are located on either side of the equator. Between them, they can see almost every part of both the northern and southern skies. The two telescopes are linked through a special high-speed internet connection.

Very Large Telescope (VLT) Array

- **Size of primary mirrors** 27 ft (8.2 m) each
- **Location** Mount Paranal, Chile
- **Altitude** 8,645 ft (2,635 m)

The VLT Array consists of four 27 ft (8.2 m) telescopes and four movable 4 ft (1.8 m) telescopes. The telescopes work together by combining the light beams from each telescope using a system of underground mirrors.

Large Binocular Telescope (LBT)

- **Size of primary mirrors** 28 ft (8.4 m) each
- **Location** Mount Graham, Arizona
- **Altitude** 10,700 ft (3,260 m)

The LBT has two 28 ft (8.4 m) primary mirrors mounted side-by-side, which collect as much light as one mirror measuring 39 ft (11.8 m) across. The LBT is currently the largest and most powerful single telescope in the world.

Hale Telescope

- **Size of primary mirror** 16 ft (5 m)
- **Location** Palomar Mountain, California
- **Altitude** 5,580 ft (1,700 m)

Even today, more than 60 years after it was built, the Hale telescope is the second-largest telescope using mirrors made of a single piece of glass. Mirrors much larger than this tend to sag under their own weight, distorting the image received.

European Extremely Large Telescope (E–ELT)

- **Size of primary mirror** 130 ft (39.3 m)
- **Location** Chile
- **Altitude** 10,040 ft (3,060 m)

This revolutionary telescope should come into operation in 2025. The primary mirror will be 130 ft (39.3 m) across and will collect 14 times more light than the largest telescopes in use today. One of its main objectives is to find Earth-like planets orbiting other stars.

Thirty Meter Telescope (TMT)

- **Size of primary mirror** 98 ft (30 m)
- **Location** Mauna Kea, Hawaii
- **Altitude** About 13,000 ft (4,000 m)

The $300 million TMT is expected to be operational by 2027. At its heart will be a primary mirror measuring 98 ft (30 m) in diameter, made up of 492 hexagonal segments. It will collect almost 10 times more light than one of the 33 ft (10 m) Keck telescopes. Astronomers will use the TMT to observe the formation and development of new galaxies.

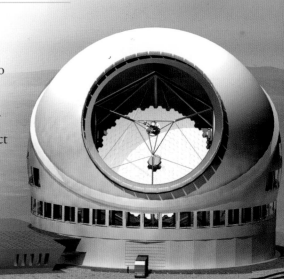

Seeing *light*

Light, the fastest thing in the Universe, is both a particle and a wave moving at about 670 million mph (just over 1 billion kph). That means it can travel from New York to London in just two-hundredths of a second—faster than the blink of an eye!

Now you see it
If you look at a beam of light, it appears to be white. However, when white light hits a shaped piece of glass called a prism, it splits into a rainbow. We call these colors, or wavelengths, of light the "visible spectrum," because our eyes can see them.

▲ WHITE LIGHT *contains a mixture of all the wavelengths of light in the visible spectrum.*

▲ WHEN A BEAM *of white light strikes the surface of a prism, it is bent. But each different wavelength is bent by a slightly different amount, and this splits the light into its spectrum of colors.*

WAVES OF ENERGY

Light carries different amounts of energy depending on its wavelength. A wavelength is the distance between the peak of one wave and the next. The higher the energy of the wave, the shorter the distance between its peaks. The complete range of waves is known as the electromagnetic spectrum.

GAMMA RAYS

▲ *Gamma rays have the highest energy and the shortest wavelengths. They are released as heat when cosmic rays collide with ordinary matter, quasars, or supermassive black holes.*

X-RAYS

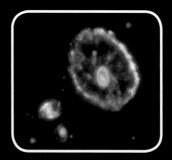

▲ *The bright white areas around the rim of the Cartwheel Galaxy are thought to be neutron stars and black holes emitting powerful X-rays.*

ULTRAVIOLET (UV)

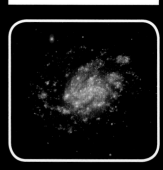

▲ *The blue areas in this image of the NGC 300 Galaxy are regions of star formation. New stars give off mainly ultraviolet light.*

▶ WE CAN USE *light to measure the composition and heat of things. This is how we know that the Boomerang Nebula is the coldest object in space at –458°F (–272°C).*

Spectroscopy

This technique uses color to determine what stars are made from and how hot they are. Every element produces its own pattern of light called a spectrum. We can see that spectrum when light passes through a special tool called a diffraction grating. By looking at the patterns, scientists can tell which elements are present and how much of that element there is.

Using the spectrum

Even though we can't see all the wavelengths, we can detect them and use them to discover things that are usually invisible. All types of matter radiate some form of energy, which means they can be picked up by telescopes that are sensitive to different parts of the electromagnetic spectrum.

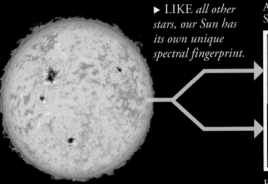

▶ LIKE *all other stars, our Sun has its own unique spectral fingerprint.*

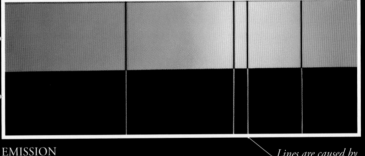

ABSORPTION SPECTRA

EMISSION SPECTRA

Absorption spectra show that some light is being absorbed before it reaches our eyes or instruments.

Emission spectra show the patterns of light emitted by specific atoms or compounds.

Lines are caused by atoms that absorb or emit radiation at specific wavelengths.

WAVELENGTH

The colors we see are all part of the visible spectrum.

VISIBLE RAYS

▲ *The Sun's visible light is only a tiny part of the energy that it radiates. Our eyes can't see the other wavelengths, but we can feel infrared heat.*

INFRARED

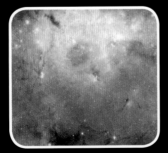

▲ *Using infrared enabled astronomers to see through the dust of the Milky Way. It revealed three baby stars that had not been seen before.*

MICROWAVES

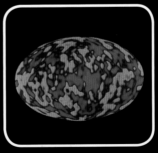

▲ *The leftover heat from the Big Bang was detected using microwaves. It is only 2.7 degrees above absolute zero, which is as cold as you can get.*

RADIO WAVES

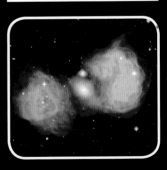

▲ *These have the longest wavelengths. The massive black hole at the center of the galaxy Fornax A is a powerful source of radio waves, shown in orange.*

Infrared *astronomy*

We are all familiar with the colors of the rainbow—red, orange, yellow, green, blue, indigo, and violet. These colors are part of what is known as the visible spectrum. Beyond the red end of the spectrum is infrared light, which we call heat. Although we cannot see infrared light, we can detect it using special telescopes, which reveal things usually hidden by clouds of dust.

SATURN'S HOT SPOT

Infrared images of Saturn reveal that it has a "hot spot"—the first warm polar cap to be discovered. This is the hottest part of Saturn and is 8–10 degrees warmer than at the equator. A huge storm thousands of miles across constantly rages over Saturn's south pole.

◄ *Infrared image of Saturn. The paler areas show where Saturn is warmest.*

A GALAXY FAR, FAR AWAY …

Messier 81 is a spiral galaxy located in the northern constellation of Ursa Major (the Great Bear). Messier 81, also known as Bode's Galaxy, is about 12 million light-years from Earth. M81 is easily visible through binoculars or small telescopes. In infrared light, the spiral arms are very noticeable because they contain dust that has been heated by hot, massive, newly born stars.

SPITZER SPACE TELESCOPE

Infrared light from space is almost completely absorbed by Earth's atmosphere, so infrared telescopes are placed on high mountains, on aircraft, or on satellites. NASA's Spitzer Space Telescope was one of the most powerful infrared observatories. Spitzer took 18 hours and over 11,000 exposures to compose this image of the Andromeda Galaxy (below).

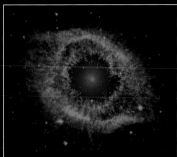

▲ THE EYE IN THE SKY
Resembling a giant eye in space, this infrared view of the Helix Nebula reveals a bright cloud of dust surrounding a dying star.

🎥 WHAT A STAR!

Frederick William Herschel (1738–1822) was a German astronomer and musician. Using a prism to split sunlight and a thermometer to detect heat, Herschel proved that there are invisible forms of light that occur beyond the visible color spectrum. This invisible heat was later called "infrared"—meaning "below red."

▲ THIS IS HOW *we usually see the Andromeda Galaxy in visible light. The main infrared image (above) has revealed its spiral arms in greater detail. Their structure is very uneven, which suggests that Andromeda may have been affected by collisions with its two satellite galaxies in the past.*

🔍 TAKE A LOOK: CONSTELLATION ORION

Looking up at the constellation Orion, you should be able to make out the stars that form its outline. You should also be able to see the bright patch of the Orion Nebula beneath Orion's Belt. This nebula is a stellar nursery where new stars are being born. If your eyes were sensitive to infrared light, then when you looked at Orion, you would see a huge dust cloud with bright patches where young stars heat the surrounding dust. The stars themselves are too hot to be seen in infrared light.

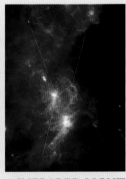

▲ VISIBLE LIGHT
The stars of the constellation Orion.

▲ INFRARED LIGHT
Bright dust clouds surrounding Orion.

Messages from the *stars*

American engineer Karl Jansky was the first to discover radio waves coming from space, using a homemade aerial in 1931. Today, scientists use radio waves to learn about all kinds of objects in space and have even attempted to contact alien life.

RADIO ASTRONOMY

Radio astronomy is the study of objects in space that produce radio waves. Radio waves are like waves of light but have far longer wavelengths than the visible spectrum. Invisible radio waves are detected by radio telescopes and can then be converted into images for us to see.

Numbers, from 1 to 10, showing how we count.

Symbols representing important chemicals found in life on Earth.

The DNA molecule—the blueprint for life on Earth.

A human form and the population of Earth.

Earth's position in the solar system.

A symbol representing the Arecibo telescope.

▲ IS THERE ANYBODY THERE?
Arecibo was used to transmit this coded message into space in 1974. So far, we haven't had a reply.

Movie star
The Arecibo telescope has featured in *Contact*, a film about first contact with extraterrestrial life, and the James Bond film *Goldeneye*.

Arecibo

One of the largest single radio telescopes in the world is Arecibo, on the Caribbean island of Puerto Rico. The telescope measures 1,000 ft (305 m) across and its dish is built into a dip in the hillside, with the radio receiver suspended 450 ft (137 m) above like a giant steel spider. Although Arecibo's dish doesn't move, its location near the equator means it can see a wide region of the sky.

TELESCOPE NETWORKS

MERLIN

- **Dish** various sizes
- **Location** various sites, UK

MERLIN is a network of seven dishes across the UK. Operated from Jodrell Bank, it includes the 250 ft (76.2 m) Lovell telescope. Altogether, the network forms a telescope equal to a single dish 135 miles (217 km) wide. It is so powerful, it can detect a coin up to 62 miles (100 km) away.

VLBA

- **Dish** 82 ft (25 m)
- **Location** Hawaii, mainland US, West Indies

The Very Long Baseline Array (VLBA) is a system of 10 radio telescope antennas. The combined effect is equal to that of a single dish more than 5,000 miles (8,000 km) wide. The VLBA can see things in such fine detail that it is equivalent to a person standing in New York reading a newspaper in Los Angeles!

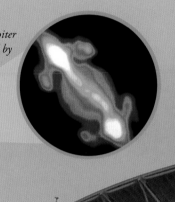

In this radio image, Jupiter is shown to be encircled by a belt of radiation.

The parabolic dish reflects the signal to the subreflector.

The subreflector focuses the signal into the receiver.

Jupiter calling Earth …
The first radio signals from a distant planet were detected from Jupiter in 1955. Since then, all of the giant gas planets have been shown to produce radio waves. Radio signals can also be bounced off the rocky planets and asteroids.

Very Large Array
One of the most important radio astronomy observatories in the world is the Very Large Array (VLA) in New Mexico. The VLA has 27 dishes arranged in a Y shape. Each arm of the Y is almost 13 miles (21 km) long. When the radio signals from each dish are combined, the whole array is equal to a giant antenna 22 miles (36 km) wide.

The 82 ft (25 m) wide dishes can be moved along tracks to change their positions.

Invisible *rays*

Ultraviolet (UV) light, X-rays, and gamma rays are types of electromagnetic radiation emitted by extremely hot objects. They are invisible and most are absorbed by Earth's atmosphere, so the best way to view them is with telescopes on high-flying balloons, rockets, or spacecraft.

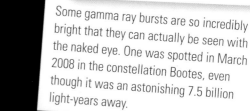

📷 **WATCH THIS SPACE**

Some gamma ray bursts are so incredibly bright that they can actually be seen with the naked eye. One was spotted in March 2008 in the constellation Bootes, even though it was an astonishing 7.5 billion light-years away.

◄ THE BALLOON *was made of thin plastic and was 360 ft (110 m) wide—big enough to fit two Boeing 767 planes inside!*

▲ THIS *telescope was lifted by a balloon into the sky over the Arctic Circle. Since the Sun never sets there in the summer, the scientists could monitor the Sun all day.*

Flying high

Although only in the air for six days, this helium balloon, part of a project called Sunrise, helped astronomers to get a unique look in UV light at how the Sun's magnetic fields form. It lifted a large solar telescope 23 miles (37 km) into the sky, high above the obscuring effects of Earth's atmosphere.

GAMMA RAY BURSTS

Gamma rays are the most energetic form of light. Gamma ray bursts, known as GRBs, are likely caused when black holes or neutron stars collide.

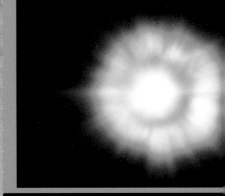

Gamma rays   *Invisible rays*

INTEGRAL

The INTEGRAL space observatory is equipped with highly sensitive detectors that can view objects in X-rays, gamma rays, and visible light all at the same time. Sent into space in 2002, it circles Earth every three days on the lookout for explosive GRBs, supernova explosions, and black holes.

▶ USING EARTH *as a shield to block emissions from distant black holes, INTEGRAL has discovered both strong and faint gamma ray and X-ray signals coming from our galaxy, possibly signals from neutron stars and black holes.*

SDO

The Solar Dynamics Observatory (SDO) studies the Sun at many different wavelengths, particularly those at the extreme end of UV. Scientists use the data that it collects from its continuous observations to learn more about how solar activity affects life on Earth.

THE X-RAY MOON

Scientists were surprised when they found that even fairly cold objects, like the Moon, can give off weak X-rays. Here, the visible Moon is compared with an X-ray image of the same area. The X-rays are produced when solar X-rays from the Sun bombard the Moon's surface and excite the atoms in the rocks.

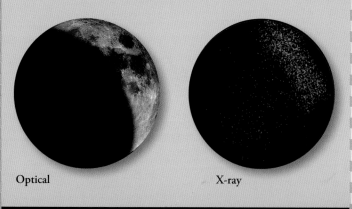

Optical

X-ray

THE SUN

With an optical telescope, we just see a scattering of dark sunspots on the Sun. When these spots are viewed with an ultraviolet-light telescope, hot, explosive solar flares can be seen.

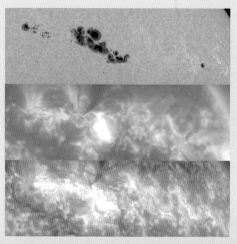

Optical

Ultraviolet

Extreme ultraviolet

X-rays

Ultraviolet (UV) rays

Visible rays

Hubble *Space* Telescope

The Hubble Space Telescope (HST) is the most famous space observatory. Since being placed in a low Earth orbit by space shuttle Discovery in April 1990, Hubble has sent back a huge amount of scientific data and incredibly detailed images of objects in space.

📷 WHAT A STAR!

Edwin Hubble (1889–1953) was the first person to prove that there are other galaxies beyond the Milky Way and that these galaxies are moving away from each other as the Universe expands.

Almost every part of HST has been replaced during its lifetime. Once repaired and upgraded, it is released back into orbit.

SERVICING MISSIONS

Hubble is the only telescope designed to be serviced in space. During service missions, space shuttles would fly alongside the telescope, take ahold of it with a robotic arm, and place it within the shuttle's cargo bay. Astronauts could then carry out repairs and replace old instruments.

Blurred vision of space

The HST mission met with a major setback when it was launched, and the first images it sent back were blurry. The cause was eventually tracked down to a mirror that had been incorrectly polished and was too flat at the edges by about one-fiftieth of the width of a human hair! The problem was finally solved 3 years later, when astronauts added lenses to correct the focus.

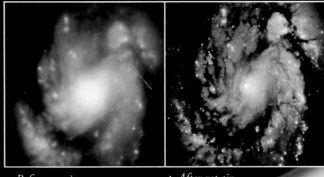

▲ *Before repair*　　　　　　▲ *After repair*

Eye on the Universe

Hubble has taken images of the Moon, Pluto, and almost every planet in the solar system—except Mercury, which sits too close to the Sun. It has also sent back amazing images of dust clouds where stars are dying and being born and provided images of thousands of galaxies. The picture on the right is of the Butterfly Nebula, a cloud of gas and dust ejected by a dying star. The image was taken by the HST's newest and most advanced camera, installed in 2009.

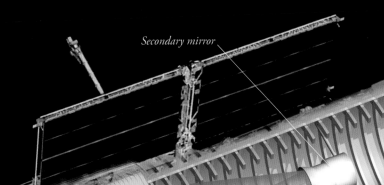

Secondary mirror

Path of light

Hubble's aperture door can be closed if it is in danger of letting light from the Sun, Earth, or Moon into the telescope.

Primary mirror: problems with the shape of the mirror were solved using corrective "eyeglasses."

Instrument module

Solar panels: power generated by the panels is also stored in six batteries and used to power Hubble when it flies through Earth's shadow.

▲ SATELLITES
Hubble communicates with the ground via NASA's Tracking and Data Relay Satellite System (TDRSS).

📷 **WHAT A STAR!**

Nancy Grace Roman (1925–2018) was instrumental in the early development of the Hubble Space Telescope, which led to NASA calling her "Mother of Hubble."

HST FAST FACTS

- **Length** 43.3 ft (13.2 m)
- **Diameter** 13.8 ft (4.2 m)
- **Weight** 24,490 lb (11,110 kg)
- **Launch date** April 24, 1990
- **Cost at launch** $1.5 billion
- **Orbit** 354 miles (569 km) above Earth
- **Speed** 17,500 mph (28,000 kph)

▲ SIGNALS *from TDRSS are received at the White Sands Ground Terminal in New Mexico.*

▲ GROUND CONTROL
Hubble is controlled from the Goddard Space Flight Center in Maryland.

WITCHES AND GIANTS

The man in the Moon may be fiction, but there is
a witch in space! The Witch Head Nebula is in the
constellation Eridanus, a safe distance of 900 light-
years from Earth. With her hooked nose and pointed
chin, she glows blue in the reflected light of Rigel,
a bright supergiant star (not seen in this picture).

▲ EYE SEE YOU *Shown in infrared light, the center of spiral galaxy NGC 1097 looks like an eye. A small companion galaxy is caught up in its arms on the left.*

▲ STAR LIGHT *Pismis 24 is an open cluster of stars. It contains three of the most massive stars ever observed. Stars are still forming in the glowing nebula (bottom).*

▲ F-ANT-ASTIC *The "body" of the Ant Nebula is actually two lobes of fiery gas ejected from a dying star at speeds of up to 600 miles (1,000 km) per second.*

▲ BUBBLE BLOWER *Young star HH 46/47 blows out two jets of warm gas. The jets have crashed into the dust and gas around the star, forming huge bubbles.*

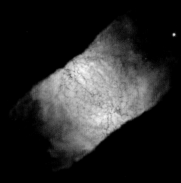

▲ SQUARE-EYED *The Retina Nebula has an unusual cylinder shape, appearing square from the side. Hot gas escapes from each end, and dust darkens the walls.*

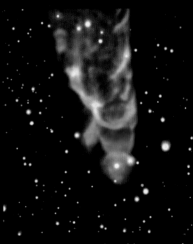

▲ JUMBO JET *Looking like a tornado in space, HH 49/50 is a jet of churned up dust and gas ejected from a young star (off the top of the picture). It is 0.3 light-years long.*

Observatories in *space*

Most of the high-energy particles and radiation emitted from objects in space are filtered out by the blanket of air around Earth. The moving atmosphere also causes shimmering or twinkling, making it hard to obtain sharp images. To study these objects, it is much easier to observe them from space observatories.

XMM-Newton can pick up faint X-rays that Chandra cannot detect.

Chandra
NASA

- **Named in honor of** the Nobel Prize–winning scientist Subrahmanyan Chandrasekhar.
- **What is it?** X-ray observatory
- **Launched** July 1999
- **Equipped with** four cylindrical mirrors nested inside each other.
- **Orbit** Circles Earth every 65 hours in an elliptical orbit 6,200–86,500 miles (10,000–139,000 km) high.

Chandra can detect X-rays from hot regions of the Universe, such as exploded stars, galaxy clusters, and the edges of black holes. It can even observe X-rays from particles just before they fall into a black hole. The first X-ray emission it saw was from the supermassive black hole at the center of the Milky Way.

▲ *Chandra flies 200 times higher than Hubble.*

XMM-Newton
European Space Agency—ESA

- **Named in honor of** the famous 17th-century scientist Sir Isaac Newton; XMM stands for X-ray Multi-Mirror.
- **What is it?** X-ray observatory
- **Launched** December 1999
- **Equipped with** three X-ray telescopes, each containing 58 concentric mirrors that are nested inside each other.
- **Orbit** Circles Earth every 48 hours in an elliptical orbit between 4,350 miles (7,000 km) and 70,800 miles (114,000 km) high.

Since X-rays pass through ordinary mirrors, X-ray telescopes are equipped with curved mirrors fitted inside each other. The X-rays glance off these mirrors and reach the detectors.

▲ *Starburst galaxy M82, the Cigar Galaxy, caught by XMM-Newton.*

TAKE A LOOK: A CLOUD OF MANY COLORS

Each space observatory highlights different aspects of celestial objects, such as Cassiopeia A, the youngest known supernova remnant in our Milky Way Galaxy. It lies about 10,000 light-years away. The rapidly expanding cloud is thought to be the remains of a massive star that exploded as a supernova around 1680.

▲ OPTICAL IMAGE FROM HUBBLE *The visible light image shows huge swirls of debris glowing with the heat generated by a shockwave from the blast.*

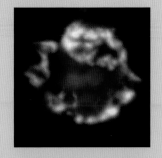

▲ INFRARED IMAGE FROM SPITZER *Hot gas (green and blue) and cool dust (red) combine in the yellow areas, showing both were created in the explosion.*

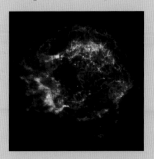

▲ X-RAY IMAGE FROM CHANDRA *The ever-expanding cloud of hot gas from the explosion is clearly visible—in fact, it is 10 light-years in diameter!*

▲ MULTICOLORED *Combining images from Hubble (yellow), Spitzer (red), and Chandra (green and blue) can help explain how supernovas evolve.*

Fermi Gamma-ray Space Telescope
NASA

- **Named in honor of** the Nobel Prize–winning Italian scientist, Enrico Fermi, a pioneer in high-energy physics.
- **What is it?** Gamma-ray observatory
- **Launched** June 2008
- **Equipped with** Large Area Telescope (LAT) and a Gamma-ray Burst Monitor (GBM).
- **Orbit** Circles Earth every 95 minutes, 340 miles (550 km) high.

This telescope was developed by the US, France, Germany, Italy, Japan, and Sweden. The satellite can turn to observe new gamma rays without commands from the ground.

▶ *This telescope has discovered many new pulsars (p.228).*

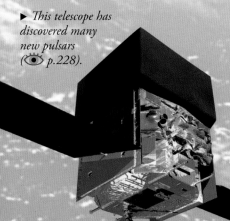

Herschel Telescope
European Space Agency—ESA

- **Named in honor of** William Herschel, the German-British astronomer who discovered infrared light and the planet Uranus.
- **What is it?** Infrared telescope
- **Launched** May 2009
- **Equipped with** 11 ft (3.5 m) wide main mirror and three supercooled science instruments.
- **Orbit** Herschel operated from an area in space located 930,000 miles (1.5 million km) from the Earth in the direction opposite from the Sun.

Able to detect a wide range of wavelengths, Herschel investigated how the first galaxies were formed and evolved and was able to probe cold, dense clouds of dust in more detail than ever before. It was deactivated in June 2013.

▲ *Instruments were supercooled using helium.*

James Webb Space Telescope
NASA

- **Named in honor of** NASA's former chief.
- **What is it?** An optical and infrared space telescope; considered to be the successor to the Hubble Space Telescope.
- **Launch date** 2021
- **Equipped with** 21¼ ft (6.5 m) primary mirror, the largest mirror ever flown in space.
- **Orbit** 932 million miles (1.5 million km) away on the night side of Earth.

The US, Europe, and Canada have all contributed to this giant telescope. Once launched, it will be able to study the farthest and faintest objects in the Universe.

Sunshield

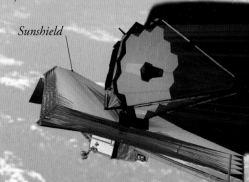

▲ *The sunshield is the size of a tennis court.*

Unusual observatories

Scientists today use all sorts of strange instruments to observe the Universe. Here are a few of the more unusual ones from around the world.

▲ *THESE unassuming white containers contain highly sensitive equipment that monitors the Sun.*

GONG
The Global Oscillation Network Group

- **Location** Six stations around the world (California, Hawaii, Australia, India, Canary Islands, and Chile).
- **Function** Studies sound waves from the Sun.

These observatories study sound waves moving inside the Sun by detecting small quakes on its surface. These quakes excite millions of sound waves, each one carrying a message about the Sun's interior.

LIGO
The Laser Interferometer Gravitational-Wave Observatory

- **Location** Three detectors in Washington and Louisiana.
- **Equipped with** L-shaped observatory with 2½ mile (4 km) long tubes containing laser beams and mirrors.
- **Function** Searches for gravitational waves.

Gravitational waves are thought to be ripples in space–time, possibly produced when black holes collide or supernovas explode. They may also have been generated in the early Universe. LIGO detected gravitational waves for the first time in 2015.

▲ *If a gravitational wave passes through Earth, it will affect the light from the laser beams in the tubes.*

South Pole Telescope (SPT)
The Arcminute Cosmology Bolometer Array Receiver

- **Location** Amundsen-Scott Research Station, South Pole.
- **Equipped with** 33 ft (10 m) telescope.
- **Function** Observes microwave background radiation.

In the Antarctic winter, sunlight doesn't reach the South Pole, so it is dark both night and day. The extremely dry air makes it a perfect location to search for tiny variations in the radiation left over from the Big Bang.

▼ *The telescope has to be supercooled to ¼ of a degree above absolute zero, −459°F (−273°C).*

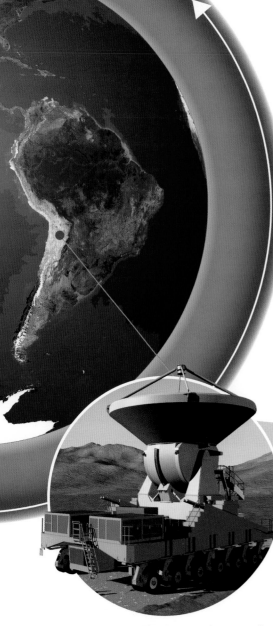

► *SOFIA 747SP is able to keep its telescope pointing steadily at an object in space even if the aircraft is struck by turbulence.*

▲ *TRANSPORTERS are used to move the giant antenna dishes to different positions.*

SOFIA
The Stratospheric Observatory for Infrared Astronomy

- **Location** The left-hand side of the fuselage of a modified Boeing 747SP.
- **Equipped with** A 100 in (2.5 m) diameter reflecting telescope.
- **Function** To observe the sky in visible and infrared light.

The SOFIA aircraft flies above the clouds and most of the atmosphere at altitudes of between 7 and 8½ miles (11 and 14 km) for up to 8 hours at a time. Since its first flight in 2010, it has made important discoveries about the atmospheres of distant worlds and the composition of gas clouds between the stars.

SNO
Sudbury Neutrino Observatory

- **Location** 6,800 ft (2 km) underground in a working nickel mine, Sudbury, Ontario, Canada.
- **Equipped with** "Heavy" water in a 39 ft (12 m) diameter tank, surrounded by 9,600 sensors.
- **Function** To study high-energy particles (neutrinos) from the Sun's core and exploding stars.

Neutrinos usually pass undetected through Earth, but when they collide with the heavy water atoms, they produce light flashes, which are picked up by the sensors surrounding the tank.

▼ *The rock shields the detectors from cosmic rays.*

ALMA
Atacama Large Millimeter/submillimeter Array

- **Location** A 16,400 ft (5,000 m) high plateau in the Atacama Desert, Chile.
- **Equipped with** At least 66 antennas across 200 pads over 11½ miles (18.5 km).
- **Function** To observe the gas and dust of the cool Universe.

ALMA is a collection of 66 dishes up to 39 ft (12 m) across that can operate together as a single giant telescope. The dry climate, together with the thin atmosphere at such a high altitude, is perfect for clear views of infrared and microwave radiation from space.

THE VIOLENT UNIVERSE

Ever-changing and full of action, the universe contains everything that exists: all matter from the smallest atom to the largest galaxy cluster, the emptiness of space, and every single second of time.

What is the *Universe?*

The Universe is everything that exists—planets, stars, galaxies, and the space between them. Even time is part of the Universe. No one knows how big the Universe is or where it starts and ends. Everything is so far away from our own little planet that light from stars and galaxies can take billions of years to reach us—so we see the distant Universe as it looked billions of years ago. But we can use the information this light provides to discover how the Universe began and how it might end.

LIGHT-YEARS

◄ WE CAN *find out what the Universe was like early in its history by using different types of telescopes.*

Telescopes are like time machines. They detect light that has traveled from distant stars and galaxies. This means that we see them as they were when the light started on its journey—thousands or even billions of years ago. Astronomers can measure the size of the Universe in light-years. A light-year is the distance light travels in 1 year—about 6 trillion miles (9.5 trillion km). Light from the farthest galaxies we can see has taken about 13 billion years to reach us. We see them today as they were long before the Sun and Earth came into existence.

Now you see it …
Light travels through empty space at 186,000 miles a second (300,000 kilometers a second). At this speed, light waves could travel around the world seven times in a single second.

FUTURE UNIVERSE

For many years, scientists believed that the pull of gravity from the stars and galaxies would gradually slow down the expansion of the Universe. However, recent observations suggest that this expansion is accelerating. If it is true, the galaxies will get farther and farther apart. No more stars will form, black holes will disappear, and the Universe will end as a cold, dark, lifeless, and empty place.

We can see and measure three
dimensions of space—height, width,
and depth. Time is a fourth dimension.
Scientists believe the Universe may
have at least six other, hidden
dimensions. These are all curled up
on each other and are infinitely tiny.

▲ AS OBJECTS *move away from
us, their light spectrum changes. By
measuring the change, we can work
out how fast they are moving.*

Measuring distances

Measuring distances in the Universe
is tricky. Many galaxies are so far
away, the only thing we can use
is light. Because the Universe is
expanding and stretching space, the
wavelengths of light from an object
also become stretched. Any dark
lines in its spectrum move toward
the red end, which astronomers call
a "redshift." By measuring the size
of this redshift, astronomers can
calculate the distances of the galaxies
and how fast they are moving away
from us. The oldest and fastest-
moving galaxies are those with
the biggest redshifts.

Shape of the Universe

Since we live inside the Universe, it is hard
to imagine that space has a shape. Scientists,
however, think that it does have a shape and
that this depends on the density of its matter.
If it is greater than a critical amount, then
the Universe is said to be closed. If it is less,
then it is described as open (saddle-shaped).
However, spacecraft observations have shown
that the Universe is very close
to the critical density, so
scientists describe it as flat.
A completely flat Universe
has no edge and will go
on expanding forever.

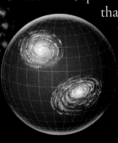

Closed

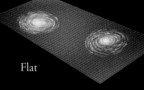

Flat

Open

◄ ALL *the stars, dust, and gas we can
see in the sky make up only a small part
of the Universe. Most of the Universe is
made of mysterious, invisible dark matter
and dark energy* (👁 *pp.62–63*).

MULTIPLE UNIVERSES?

Is our Universe alone, or are there other Universes
that we cannot see? No one knows, but some scientists
believe that there might be many other Universes. This
structure may resemble an enormous bubbly foam in
which some Universes have not yet inflated. Some may
have different physical laws and dimensions to ours. In
theory, it may even be possible to connect one Universe
to another through a spinning black hole. However,
no other Universes can affect anything in our Universe,
so it is impossible to prove that they exist.

Birth of the *Universe*

Scientists believe that the Universe was born in a huge fireball about 13.8 billion years ago. This "Big Bang" was the beginning of everything: time and space, as well as all the matter and energy in the Universe.

INFLATION

At the instant it began, the newborn Universe was incredibly small and unimaginably hot and dense. Inside the fireball, energy was being turned into matter and antimatter. Then it began to expand and cool. For a tiny fraction of a second, the expansion was quite slow, but then the Universe shot outward. It has been expanding steadily ever since and might even be speeding up.

1 The Universe begins to expand from infinitely tiny to the size of a grapefruit. The huge amount of energy this releases kick-starts the formation of matter and antimatter.

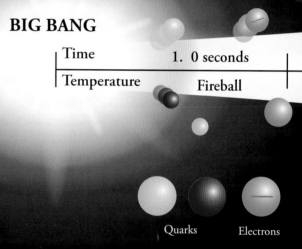

BIG BANG

Time	1. 0 seconds
Temperature	Fireball

▼ THE BLUE AND PURPLE *colors show X-rays being given off by matter and antimatter collisions as high-energy particles stream away from the white pulsar at the center of the image.*

Quarks   Electrons

▲ THE MOST COMMON *particles in the Universe today include quarks and electrons. They are the building blocks of all atoms.*

Matter and antimatter

Immediately after the Big Bang, huge amounts of energy were turned into particles of matter and mirror-image particles of antimatter. When the two types meet, they destroy each other in a flash of radiation. If equal numbers of both had been created, they would have wiped each other out. However, everything we can see in the Universe today consists mainly of matter. The only explanation seems to be that, for some unknown reason, the Big Bang created slightly more matter than antimatter.

Which came first?

There was no "before" the Big Bang because time and space did not exist. After the Big Bang, space began to expand and time began to flow. But neither could start until the other one began. It took scientists years to figure out this mind-boggling fact!

FIRST THREE MINUTES

During the first 3 minutes, the Universe cooled from being unbelievably hot to less than 1 billion degrees Kelvin. In the same period, it expanded from an area billions of times smaller than an atom to the size of our Milky Way Galaxy.

2 By now, the Universe is the size of a football field. Huge numbers of matter and antimatter particles collide and destroy each other, creating more energy.

3 The Universe suddenly inflates and starts to cool. A new range of exotic particles form, including quarks and electrons.

4 The Universe is still too hot to form atoms, but quarks begin to group together and form heavier particles, particularly protons and neutrons.

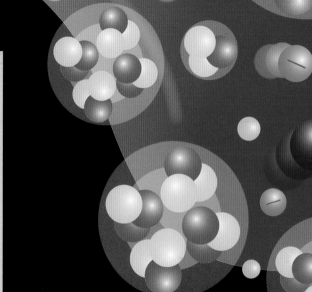

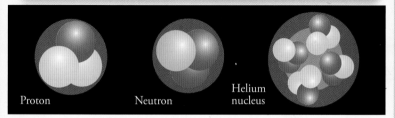

2. 10^{-43} seconds	3. 10^{-35} seconds	4. 10^{-7} seconds	3 minutes
10^{32} K	10^{27} K	10^{14} K	10^{8} K

▲ *K stands for Kelvin, a temperature scale used by astronomers; 0 K equals –459°F (–273°C). It is the lowest possible temperature anything in the Universe can reach.*

BUILDING UP TO ATOMS

Proton   Neutron   Helium nucleus

Protons and neutrons are particles that each contain three quarks. Once the expanding Universe had enough protons and neutrons, they began to form very simple atomic nuclei, the basis of hydrogen and helium atoms. Most stars are made of these two types of atoms. Within 3 minutes of the Big Bang, almost all of the hydrogen and helium nuclei in the Universe had been created.

It took hundreds of millions of years for stars, galaxies, and planets to start filling the Universe. If the Universe hadn't begun to cool, the atoms they are made from would never have formed.

THE FOGGY UNIVERSE

Around 300,000 years passed before the first atoms started to form. This process began when the temperature of the Universe dropped to about 3,000 K. In this cooler Universe, protons and atomic nuclei were able to capture extremely tiny particles called electrons and become atoms. Until this point, the Universe was very foggy—light could not travel far because it was constantly bouncing off atomic particles. This fog is why we cannot see anything that was happening at that time—even with the most powerful telescopes.

300,000 years

3,000 K

WHAT IS AN ATOM?

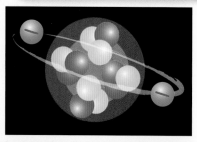

An atom is the smallest piece of matter that can exist on its own. Atoms have a central core (the nucleus) made up of protons and neutrons. Circling the nucleus are electrons. The number of protons, neutrons, and electrons an atom has determines which element it is. When the first stars exploded as supernovas, the energy they released created new, heavier elements, such as carbon, oxygen, and iron. This process continues today.

FIRST STARLIGHT

About 200 million years after the Big Bang, huge clouds of hydrogen and helium gas began to build up. The pull of gravity made the clouds collapse into dense clumps of atoms. As these clouds shrank and became hotter, they ignited and formed the first stars. These stars didn't last long before they exploded and helped produce new stars.

BEGINNINGS OF GALAXIES

Galaxies also began to form fairly soon after the first stars. Dense clouds of gas and young stars were pulled together by gravity and dark matter to form small galaxies and new stars. Gradually these galaxies began colliding with each other to make larger galaxies.

FUNDAMENTAL FORCES

The Big Bang also created four fundamental forces that affect the Universe. These are gravity, the electromagnetic force, the weak nuclear force, and the strong nuclear force. Gravity is what keeps planets in orbit around stars. Electromagnetism is linked to electricity and magnetism. The weak force governs how stars shine, while the strong force holds together the protons and neutrons in the nuclei of atoms.

The Moon is held in orbit around Earth by the pull of gravity.

200 million years	500 million years	Present day
100 K	10 K	2.7 K

▲ THE CMB *provides the best evidence for the Big Bang. It marks the point at which the temperature dropped enough for atoms to form.*

Glowing embers of the Big Bang

We cannot see any light from the Big Bang. However, we can detect a faint glow of radiation—known as the Cosmic Microwave Background (CMB)—that still covers the sky. This leftover radiation shows what the Universe was like 300,000 years after it began. The map shows slightly warmer and cooler ripples. The first galaxies probably grew from the slightly cooler and denser (blue) patches of gas.

THE BIG BANG MACHINE

Scientists cannot see what the Universe was like immediately after the Big Bang, but they are trying to learn more by building huge machines on Earth. The latest and most advanced of these is the Large Hadron Collider in Switzerland. This $4 billion instrument, which began operation in 2010, is attempting to recreate the Big Bang by crashing beams of protons together 1 billion times a second. The beams that collide are expected to create many new particles and possibly provide a reconstruction of the Universe in its very first moments.

100 billion *galaxies*

Wherever we look in the sky, the Universe is full of galaxies—huge star systems that are tied together by gravity. The first galaxies began to form less than 1 billion years after the birth of the Universe in the Big Bang.

GIANTS AND DWARFS

There are at least 100 billion galaxies in the Universe. Some are enormous, containing hundreds of billions of stars. Others are much smaller, sometimes containing fewer than a million stars. There are many more small galaxies than giant galaxies, even though the dwarf galaxies tend to be swallowed by their larger neighbors over time. We live in a galaxy of about 100 billion stars called the Milky Way.

▲ M51, *the Whirlpool Galaxy, is about 30 million light-years from Earth.*

M51 GALAXY

Seeing the light

There are many features of galaxies that do not show up in visible light. To find out the true nature of a galaxy, you have to look at it at different wavelengths with different instruments. The above image of M51 combines images taken by four space telescopes. One showed X-rays given off by black holes, neutron stars, and the glow from hot gas between the stars (shown in purple). Infrared and optical instruments revealed stars, gas, and dust in the spiral arms (in red and green). Young, hot stars that produce lots of ultraviolet light are blue.

🔍 TAKE A LOOK: WHIRLPOOL

By the mid-19th century, astronomers had discovered many fuzzy patches in the night sky, which they called nebulas. To find out more about them, Lord Rosse built what was then the world's largest telescope—the 72 in (1.8 m) Birr telescope. With it, he made the first observation of what is now known as the Whirlpool Galaxy (M51). His drawing of the galaxy is dated 1845.

This Hubble Space Telescope image is of Zwicky 18, a dwarf galaxy about 60 million light-years away.

ZWICKY 18

GAS GALAXIES

Some galaxies are very large yet contain very few stars. These faint galaxies are made almost entirely of gas, so in photos they appear as a smudge in the sky. One example, Malin 1, contains enough gas to make 1,000 galaxies like the Milky Way. It seems to have just begun to make stars. Its vast but faint disk is six times bigger than the Milky Way. A much closer, fully formed galaxy can be seen at the bottom of the picture.

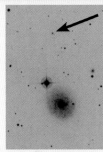 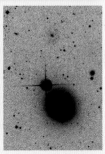

The arrow points to Malin 1.

It can be seen better in this treated image.

ULTRA-DEEP FIELD
By focussing its cameras on a square of apparently empty space in the southern constellation Fornax for a total of 11 days and 8 hours, the Hubble Space Telescope captured light from more than 10,000 distant galaxies stretching away toward the edge of the observable Universe.

Galaxy formation

Galaxies have existed for many billions of years—but where did they come from? Astronomers today use observatories to look back to the very early universe. These distant views show fuzzy galaxies involved in violent collisions. Could this be how the first galaxies formed?

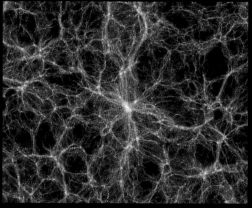

▲ THEORY TEST *This computer model shows matter clumping into strands under the influence of gravity. The first galaxies form inside these strands.*

WHAT HAPPENS?

There are two main theories of how galaxies form. In one version, huge clouds of gas and dust collapse to form galaxies. In the other version, stars form into small groups and then merge to form larger groups, then galaxies, and finally clusters of galaxies.

Changing shape

Many galaxies begin life as small spirals before becoming larger ellipticals, often as the result of a collision. This doesn't mean that the galaxies crash into each other—the gaps between the stars in a galaxy are large enough for the galaxies to pass through each other. However, it does change the galaxy's shape.

▲ YOUNG SPIRAL *NGC 300 is a young spiral galaxy with lots of star formation.*

▲ TEENAGE TRANSITION *As the galaxy grows older, there is less star formation.*

▲ OLD ELLIPTICAL *Large, gas-poor elliptical galaxies contain old stars.*

Odd one out

Hoag's Object is a very unusual galaxy. It does not look like other irregular, spiral, or elliptical galaxies. Instead, it has a circle of young blue stars surrounding its yellow nucleus (core) of older stars.

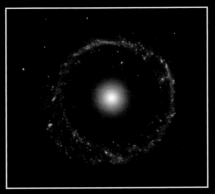

▲ BLUE RING *Clusters of hot blue stars dominate the ring. They may be the remains of another galaxy that came too close.*

◄ SMOKIN'! *The Cigar Galaxy is an irregular galaxy with a lot of star formation. More stars are formed in young galaxies than in older ones.*

TYPES OF GALAXY

There are three main types of galaxy. These are classified according to their shape and the arrangement of stars inside them.

■ **Irregular** galaxies contain a lot of gas, dust, and hot blue stars, but have no particular shape. They are often the result of a collision between two galaxies.

■ **Elliptical** galaxies are round, oval, or cigar-shaped collections of stars. They usually contain very old red and yellow stars with little dust or gas between them.

■ **Spiral** galaxies are huge, flattened disks of gas and dust that have trailing arms.

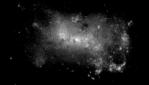

▲ IRREGULAR *These galaxies sometimes have the beginnings of spiral arms.*

▲ ELLIPTICAL *There is no gas in an elliptical galaxy so no new stars can form.*

▲ SPIRAL *Spirals rotate very slowly, about once every few hundred million years.*

STARTING A SPIRAL

Most scientists believe that the early universe was filled with hydrogen and helium. Some suggest that clouds of gas and dust, collapsing and rotating under the influence of gravity, formed spiral galaxies.

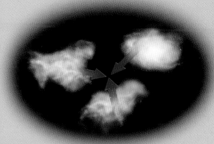

▲ COME TOGETHER *Clouds of dust, gas, and stars are pulled together by gravity.*

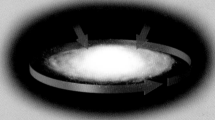

▲ TURN AROUND *Gravity makes the collapsed clouds rotate. New stars form and rotate around the center of the mass.*

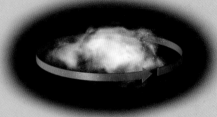

▲ SHRINK DOWN *The spinning action flattens the cloud, forming a galactic disk of dust, gas, and stars.*

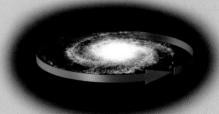

▲ TAKING UP ARMS *The disk continues to rotate, causing spiral arms to form.*

47

A SOMBRERO IN SPACE

Around 28 million light-years from Earth,
in the constellation Virgo, lies a spiral galaxy
with a very bright nucleus. It has an unusually
large central bulge and is surrounded by a
dark, inclined lane of dust (shown here in
a side-on view). Named for its hatlike
appearance, this is the Sombrero Galaxy.

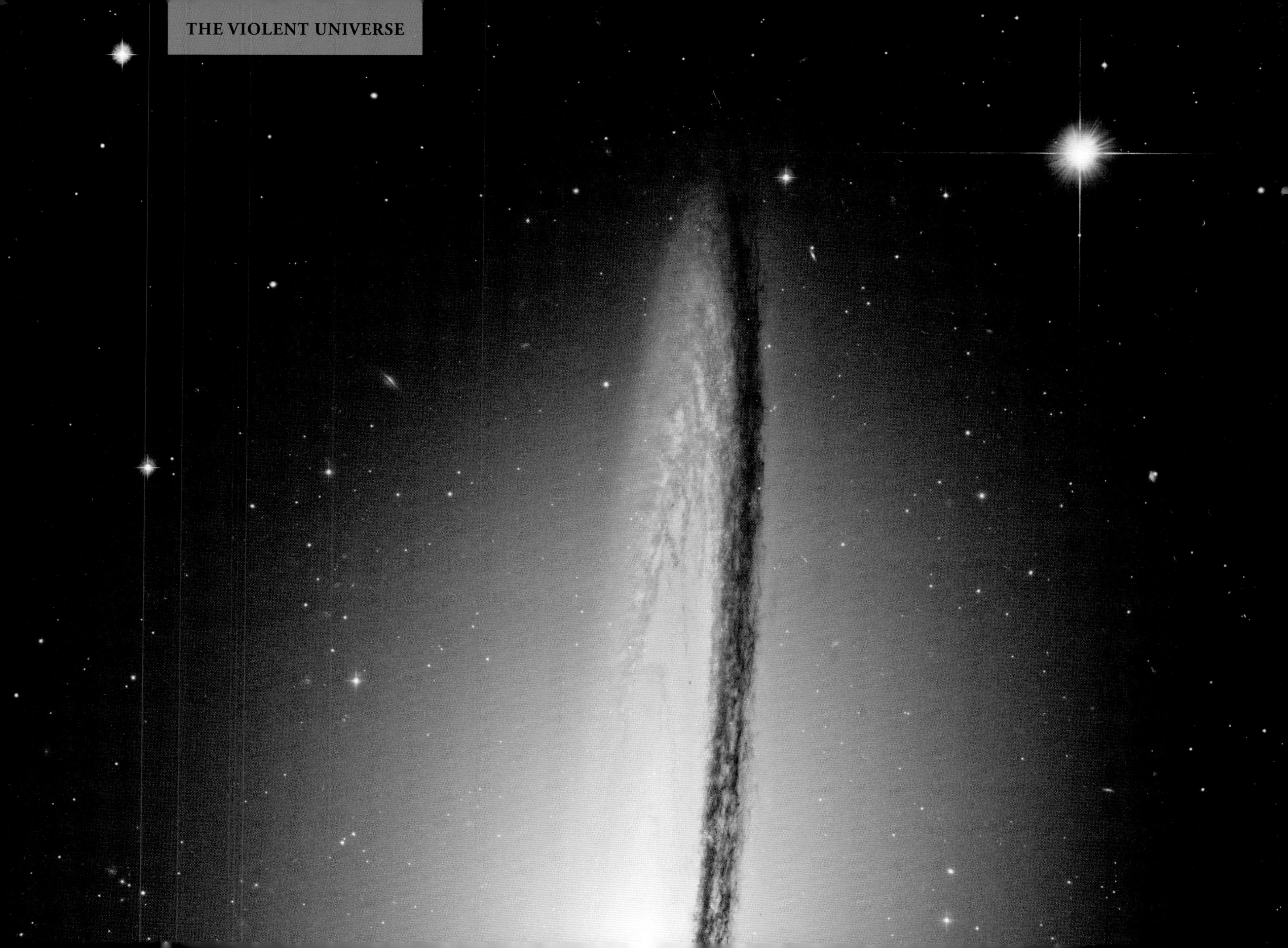

The Milky Way

We live on a small planet that circles an insignificant star in a tiny part of a huge, spiral star system—the Milky Way galaxy. The Milky Way was born more than 10 billion years ago and is likely to exist for many more billions of years.

Norma arm

Carina-Sagittarius arm

Crux-Scutum arm

Galactic center

Galactic bar

Perseus arm

Orion arm

Our Sun

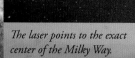

The laser points to the exact center of the Milky Way.

Seeing stars

If you live far away from bright city lights, you may be lucky enough to see a faint band of light that crosses the night sky. Ancient observers called it the Milky Way because it looked like a stream of spilled milk in the sky. They had no idea what it was, but the puzzle was solved in 1610 when Galileo turned his telescope on the Milky Way and discovered that it was made up of thousands of stars.

Solar system

Globular cluster of millions of stars

Central bulge

Dark halo

Galactic disk

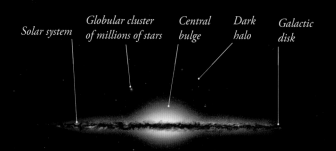

▲ HOW BIG IS OUR GALAXY? *The Milky Way is about 100,000 light-years across but only 2,000 light-years thick toward its outer edge. Most of the Milky Way's mass seems to come from mysterious, invisible dark matter (◉ pp. 62–63).*

A SPIRAL GALAXY

The Milky Way is a barred spiral galaxy, which means it is shaped like a giant pinwheel, with curved arms trailing behind as it turns. The stars in our galaxy all move around the center as the galaxy spins. Our Sun, which is about 28,000 light-years from the center, goes around the galaxy once every 220 million years. Stars near the center take less time to orbit than the Sun.

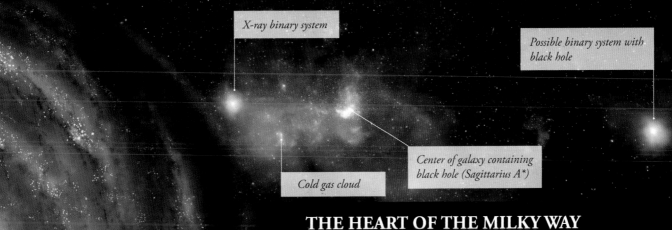

X-ray binary system

Possible binary system with black hole

Cold gas cloud

Center of galaxy containing black hole (Sagittarius A*)

The Sun is just one of about 200 billion stars in the Milky Way. Most stars lie in the galaxy's central bulge, but younger stars and dust clouds are found in the five spiral arms. A supermassive black hole lies at the center.

THE HEART OF THE MILKY WAY

The center of the Milky Way is a mysterious place about 600 light-years across. While this is just a tiny part of the galaxy, the core contains one-tenth of all the gas in the galaxy, along with billions of stars. These include the remains of supernovas and bright sources of X-rays, such as binary systems (pairs of objects) that are thought to contain a black hole.

SGR A*

▲ ACTIVE PAST *SGR A* seems to have been active in the past. Light echoes from an outburst of X-rays 300 years ago can be seen passing through nearby dust clouds.*

The hidden monster

At the center of our galaxy lies a monster: a supermassive black hole that contains about four million times more material than our Sun. This is Sagittarius A* (or SGR A*), named after its location in the constellation Sagittarius. At the moment, it is a sleeping giant, creating billions of times less energy than giant black holes in other galaxies.

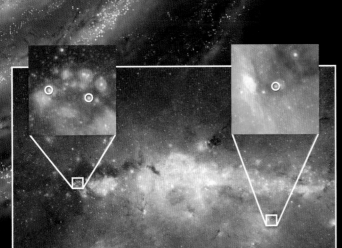

Baby stars

The heart of our galaxy is cluttered with stars, dust, and gas surrounding the black hole. Conditions there are harsh, with fierce stellar winds—powerful shock waves that make it difficult for stars to form. We don't yet know how stars form there because, until recently, no one could peer through the dust to find newborn stars. In 2009, however, the Spitzer Infrared Observatory found three baby stars, all less than one million years old, embedded in cocoons of gas and dust.

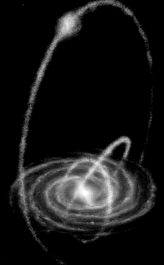

Ancient star streams

Not all of the material in the Milky Way lies in a flat disk. Three narrow streams of stars have been found arcing high above the galaxy. They are between 13,000 and 130,000 light-years from Earth and extend over much of the northern sky. The largest stream is thought to be the scattered remains of a dwarf galaxy that collided with the Milky Way.

The Magellanic *Clouds*

The Milky Way is not the only galaxy visible in our skies. In the southern hemisphere, you can also see the two Magellanic Clouds. They are generally thought to be satellite galaxies orbiting the Milky Way. Recent research indicates that the larger cloud will one day collide with the Milky Way.

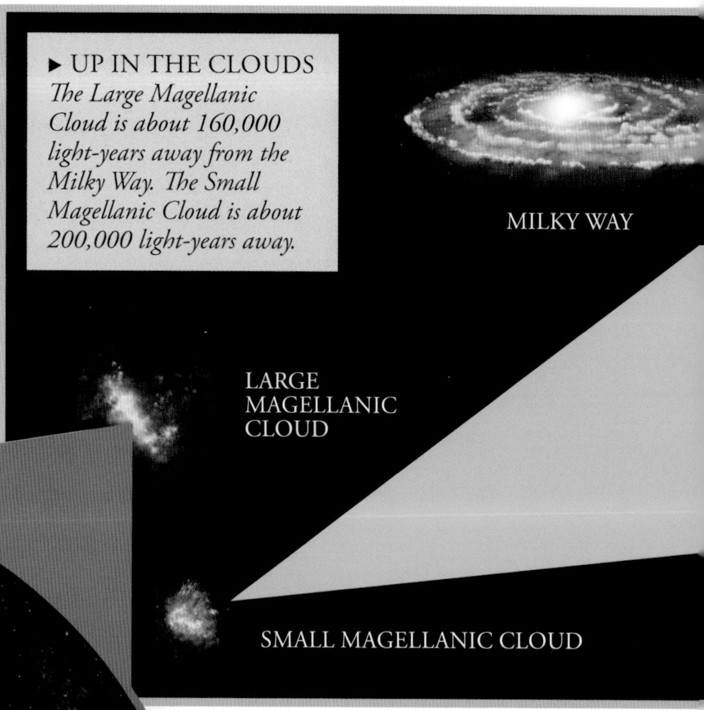

▶ **UP IN THE CLOUDS**
The Large Magellanic Cloud is about 160,000 light-years away from the Milky Way. The Small Magellanic Cloud is about 200,000 light-years away.

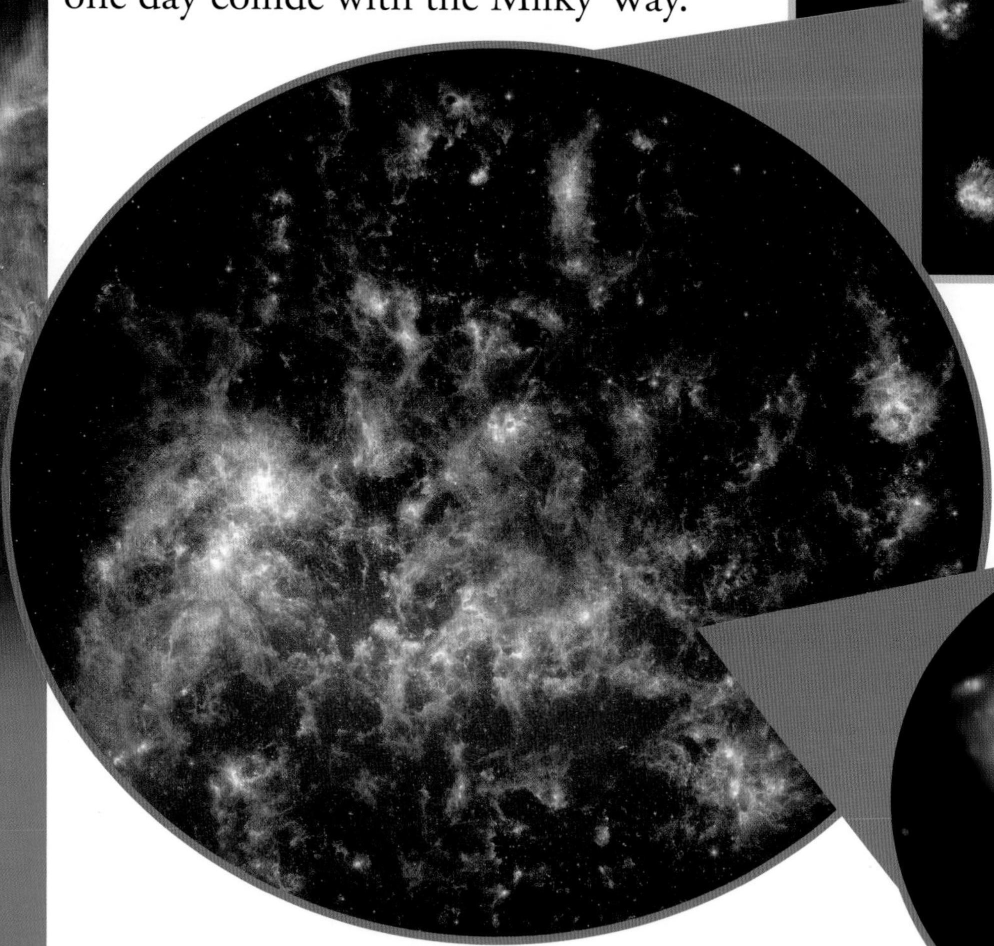

MILKY WAY

LARGE MAGELLANIC CLOUD

SMALL MAGELLANIC CLOUD

◀ **LMC CLOSE-UP** *Nearly 1 million objects are revealed in this detailed view from the Spitzer Infrared Observatory, which shows about one-third of the whole galaxy. Blue represents starlight from older stars. Red is from dust heated by stars.*

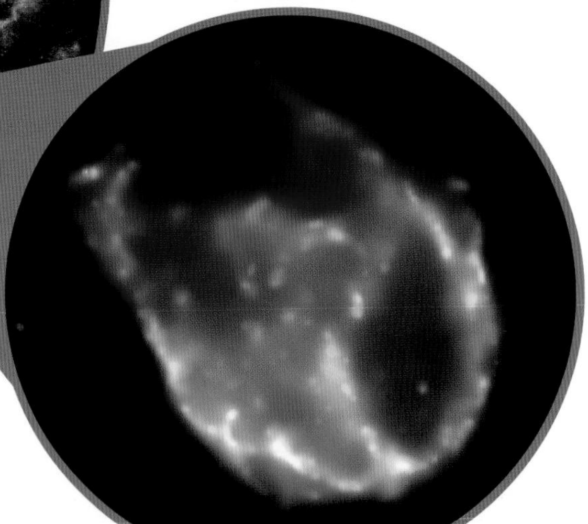

LARGE MAGELLANIC CLOUD

The Large Magellanic Cloud (LMC) lies in the southern constellations Dorado and Mensa. It is about 20,000 light-years across and contains the mass of about 100 billion Suns. The LMC is classed as an irregular galaxy, although it has a bar at its center and some signs of spiral arms. It may have once been a spiral galaxy that was pulled into a new shape by the gravity of the Milky Way.

Colorful clouds

The Magellanic Clouds contain many supernova remnants. These are the remains of massive stars that exploded thousands of years ago, leaving behind colorful expanding clouds of hot gas.

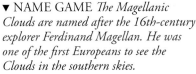

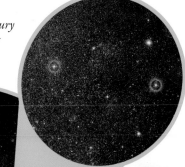

▼ NAME GAME *The Magellanic Clouds are named after the 16th-century explorer Ferdinand Magellan. He was one of the first Europeans to see the Clouds in the southern skies.*

▶ STAR NURSERY *This false-color image shows a part of the Tarantula Nebula near the star cluster NGC 2074. It shows a "nursery" where new stars form. The area has dramatic ridges, dust valleys, and streams of gas that glow in ultraviolet light.*

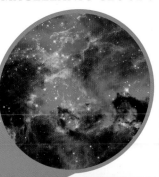

Small Magellanic Cloud

The Small Magellanic Cloud (SMC) is one of the most distant objects that can be seen with the naked eye. This irregular dwarf galaxy is a smaller version of the LMC. It contains less dust and gas, but it still has a number of star-forming regions (the red regions shown above). The SMC has a visible diameter of about 9,000 light-years and contains several hundred million stars. Its mass is about 7 billion times the mass of our Sun.

Tarantula Nebula

30 Doradus is a vast star-forming region in the LMC. The region's spidery appearance gives it its popular name, the Tarantula Nebula. It is about 1,000 light-years across and 160,000 light-years away. If it were as close as the nearest star nursery to Earth (the Orion Nebula, 1,500 light-years away), it would be visible during the day and cover a quarter of the sky. The nebula contains very hot stars that are among the most massive stars we know.

TAKE A LOOK: MAGELLANIC STREAM

The Magellanic Clouds and the Milky Way are connected by an unusual, extended ribbon of hydrogen gas—the Magellanic Stream. Visible only at radio wavelengths, the Stream extends more than halfway around the Milky Way. It may have been created when material was stripped off these galaxies as they passed through the halo of our Milky Way. Another theory suggests that the two Clouds passed close to each other, triggering massive bursts of star formation. The strong stellar winds and supernova explosions from that burst of star formation could have blown out the gas and started it flowing toward the Milky Way.

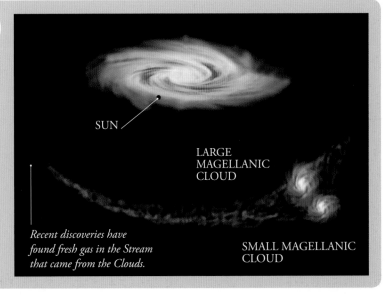

SUN

LARGE MAGELLANIC CLOUD

SMALL MAGELLANIC CLOUD

Recent discoveries have found fresh gas in the Stream that came from the Clouds.

The Local *Group*

The Milky Way is not alone in space; it is a member of a cluster of galaxies called the Local Group. The Local Group contains over 50 galaxies plus several more lying on its borders.

ANDROMEDA

The Andromeda Galaxy (M31) is our largest galactic neighbor and is about 50 percent larger in diameter than our Milky Way. The entire disk of the spiral galaxy spans about 220,000 light-years, which means that it would take 220,000 years for a light beam to travel from one end of the galaxy to the other.

OUR NEIGHBORS

The galaxies in the Local Group all lie less than 3 million light-years from the Milky Way. They are arranged into two smaller groups based around the two largest galaxies: the Milky Way and Andromeda. Astronomers think that, in several billion years, the Milky Way and Andromeda will collide and merge to form one huge elliptical galaxy.

▶ GROUPED TOGETHER
Some of the largest galaxies in the Local Group are shown here.

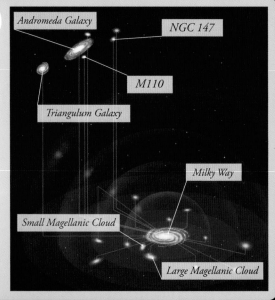

Andromeda Galaxy

NGC 147

M110

Triangulum Galaxy

Milky Way

Small Magellanic Cloud

Large Magellanic Cloud

Hot-hearted Andromeda

In the middle of Andromeda is a cloud of hot gas that gives out X-rays. The X-rays are thought to come from a binary system (a pair of stars) that contains a neutron star or a black hole that is pulling material away from a normal star. As matter falls toward the neutron star or black hole, friction heats it up to tens of millions of degrees and produces X-rays.

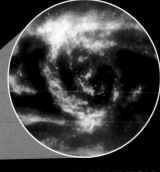

▶ CHANDRA'S VIEW
This image from NASA's Chandra X-ray Observatory shows the center of Andromeda. Low-energy X-rays are red, medium-energy X-rays are green, and blue indicates high-energy X-rays.

▲ ANCIENT COLLISION
Dust rings inside Andromeda provide evidence that the galaxy was involved in a violent head-on collision with the dwarf galaxy Messier 32 (M32) more than 200 million years ago.

 WATCH THIS SPACE

This ultraviolet and infrared image of M33 shows a mix of dust and young stars in the galaxy. In some of the outer regions of the galaxy, there are many young stars (glowing blue) and very little dust.

Dwarf galaxies

The Local Group contains several dozen dwarf galaxies and probably many more that are waiting to be discovered. Most are very small and faint, containing up to a few hundred million stars. Lurking behind dust and stars near the plane of the Milky Way is the closest known starburst galaxy—an irregular dwarf galaxy known as IC 10. Although its light is dimmed by dust, you can see the red glow of the star-forming regions.

Triangulum Galaxy

M33, or the Triangulum Galaxy, is the third largest galaxy in the Local Group. It is also known as the Pinwheel Galaxy because of its face-on spiral shape, which is more than 50,000 light-years wide. M33 is thought to be a satellite of the Andromeda Galaxy. Like Andromeda, M33 is used as a cosmic ruler for establishing the distance scale of the Universe.

THE HEART OF THE MILKY WAY

A look at the center of our galaxy reveals hundreds of thousands of stars packed into an area of sky the width of a full Moon. Near-infrared light (yellow) shows regions where stars are being born. Infrared light (red) reveals dust clouds, while X-rays (blue) show ultra-hot gas and emissions from black holes.

▼ BRILLIANT BINARY *This binary star is a major source of X-rays. It is probably a massive star being orbited by either a neutron star or a black hole.*

▲ PISTOL STAR *One of the brightest known stars in the Milky Way, it may be 10 million times brighter than the Sun.*

▲ SAGITTARIUS A*
This supermassive black hole is the center of our galaxy. Its eruptions in the past have cleared the surrounding area of gas.

are millions of light-years apart. However, some galaxies are close enough to be pulled by gravity into clusters. Members of galaxy clusters can pull on each other so strongly that they collide.

Stephan's Quintet is a group of galaxies that appear to be smashing into each other. Four of them are about 280 million light-years away from Earth, but the fifth is closer to us. NGC 7318b is passing through the main group at nearly 200 million mph (320 million km/h). This creates a shock wave that causes the gas between the galaxies to heat up and give out X-rays (the light blue region in the middle).

The NGC 7319 spiral galaxy contains a quasar (👁 *pp. 60–61*).

NGC 7318a (right) is in front of NGC 7318b (left).

NGC 7320 is much nearer to Earth than the other galaxies.

COLLISION COURSE

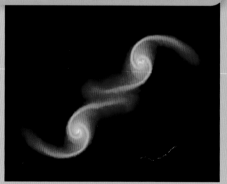

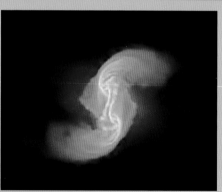

▲ VIRTUAL COLLISION *In real life, galaxy collisions take billions of years, so computers are used to see what may happen.*

▲ 6 BILLION YEARS *Since the spiral galaxies first met, gravity has begun to pull long tails from the galaxies.*

▲ 24 BILLION YEARS *In the time gap, the galaxies had separated again... until they rejoin as one slices through the other.*

This image shows gas temperature. Red is coolest, blue is hottest.

Cluster collision

The ultimate crashes occur when several clusters of galaxies collide. The biggest collision astronomers have seen so far is a pile-up of four clusters called MACS J0717. This filament (stream) of galaxies, gas, and dark matter is 13 million light-years long. It is moving into an area already packed with matter, causing repeated collisions. When the gas in two or more clusters collides, the hot gas slows down. Galaxies don't slow down as much, so they end up moving ahead of the gas.

A distorted view

Some galaxy clusters act as magnifying glasses in the sky. Their powerful gravity distorts the space around them. This means that light from more distant galaxies or quasars is bent on its way to us. We see multiple arcs and distorted images of the distant object, like a mirage in space.

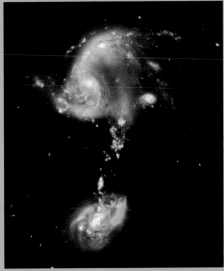

▲ ARP 194 *The top part of group ARP 194 contains two galaxies that are in the process of merging (top left in the image). The blue "fountain" running down looks as if it connects to a third galaxy, but this galaxy is much farther away and not connected at all. The fountain contains stars, gas, and dust.*

▲ THE MICE *Named after their long "tails" of stars and gas, the two interacting galaxies known as The Mice (officially called NGC 4676) will eventually join together to form one huge single galaxy. The Mice are 300 million light-years away from Earth, in the constellation Coma Berenices.*

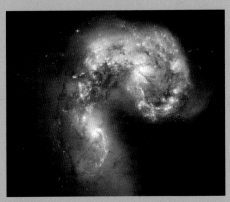

▲ THE ANTENNAE *This is the nearest and youngest pair of colliding galaxies. Early photos showed them to look like insect antennae. These "tails" were formed when the two spiral galaxies first met around 200–300 million years ago. Billions of new stars will be born as the galaxies continue to collide.*

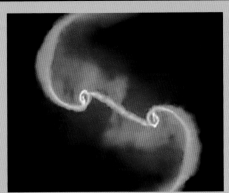

▲ 26 BILLION YEARS *The central regions fall together and the two galaxies eventually join together.*

▲ 30 BILLION YEARS *The two spiral galaxies finally merge and form one massive, elliptical galaxy.*

Active galaxi

There are many active galaxies in the univ
While our own is quiet at present, others a
generating huge amounts of energy. In the
of each is a supermassive black hole with a
gravitational pull. This is the galaxy's power

Powerful magnetic field drives high-speed jets away from the black hole.

The disk of hot gas sends out radiation such as X-rays.

Active types

There are four main types of ac
galaxy: radio galaxies, Seyfert
galaxies, blazars, and quasars (sh
for quasi-stellar objects). Radio
galaxies (such as Cygnus A show
above) are the source of the stro
radio waves in the universe. Rad
galaxies appear all over the unive
but blazars and quasars are found
only billions of light-years away.

SPINNING WHEEL

An active galaxy is like a wheel. At the hub is a black
hole. Its gravity pulls in dust, stars, and gas, making
a spinning disk with an outer "tire" of dust and gas.
A strong magnetic field around the black hole blasts
out jets of particles, looking like an axle for the wheel.

Dusty radio

The nearest radio galaxy to
Earth is Centaurus A. The
central regions of this elliptical
galaxy are hidden behind an
unusual dark, thick band of
dust. It was one of the first
objects outside the Milky Way
to be recorded as a source of
radio waves, X-rays, and
gamma rays. The two huge
plumes of radio signals (in pale
blue) are 200 million light-years
long. They were created by a
collision with a spiral galaxy.

WATCH THIS SPACE

This image of the elliptical radio galaxy
M87, taken with the Hubble Space
Telescope, reveals a brilliant jet of high-
speed electrons sent out from the nucleus.
The jet is produced by a black hole with
the mass of nearly 6 billion Suns.

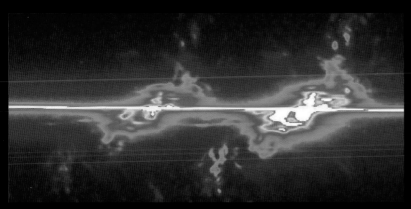

Seyfert galaxies

A Seyfert galaxy is powered by a central black hole, hundreds of millions of times the mass of the Sun. Trapped material spirals into the hole, and jets are created where some of the material is blasted out at high speed. This image of NGC 4151, the brightest Seyfert galaxy, shows a side-on view of the jets being blasted into space.

Spiraling Seyfert

M106 looks like a typical spiral galaxy, with two bright spiral arms and dark dust lanes near its nucleus. However, in radio and X-ray images, two additional spiral arms of gas can be seen between the main arms. The core of M106 also glows brightly in radio waves and X-rays, and twin jets have been found running the length of the galaxy. M106 is one of the closest examples of a Seyfert galaxy, powered by vast amounts of hot gas falling into a central massive black hole.

TAKE A LOOK: BLAZARS

A blazar is built around a supermassive black hole in a host galaxy, but the amount of energy it gives out changes over time. Our view of a blazar is different from the other active galaxies. From Earth we look down on the jets and disk, just like looking at a hole in a ring doughnut.

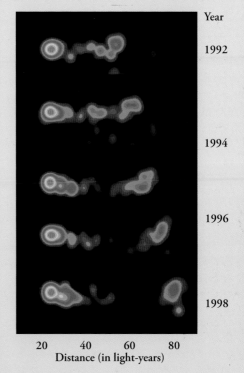

Year

1992

1994

1996

1998

20   40   60   80
Distance (in light-years)

▲ THIS SET *of images shows the movement of matter given out by blazar 3C 279. It seems to move faster than the speed of light, but this is an illusion.*

▲ COLOR CODED
In this image of M106, the gold color is what you can see in visible light. Red is the infrared view, blue is X-ray, and purple is radio waves.

Quasars

Quasars are the brilliant cores of faraway galaxies. They are similar to Seyfert galaxies, but much brighter—so bright that their light hides the fainter galaxy around them. Quasars are powered by supermassive black holes fueled by interstellar gas sucked inside. They can generate enough energy to outshine the Sun a trillion times.

Dark matter

Dark matter is the universe's biggest mystery. Astronomers can tell that there is something invisible in the spaces between stars, since it's creating enough of a gravitational pull to bend starlight as it travels toward Earth. However, no one knows what dark matter looks like or what it is made from.

▲ MISSING PIECES *At the moment, we know next to nothing about dark matter, but scientists are looking for subatomic particles that might help us complete our picture of the universe.*

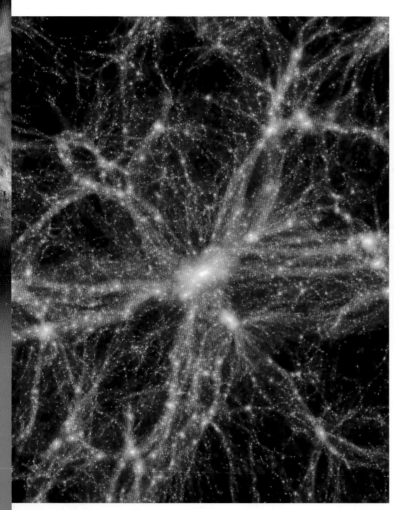

MAPPING IT OUT

This computer simulation shows how dark matter is spread throughout the universe. The yellow areas show the highest concentrations of dark matter. These regions have enough gravity to pull together visible matter, creating galaxies.

IT'S A MYSTERY

Five percent of the visible universe of stars and planets is normal matter. However, this matter would not have enough gravitational pull to hold the galaxies together, so

ATOM

astronomers know that there must be another kind of matter, even if it's invisible. Dark matter isn't made of atoms and does not reflect light or any other kind of radiation, but it appears to make up a quarter of the matter in the universe.

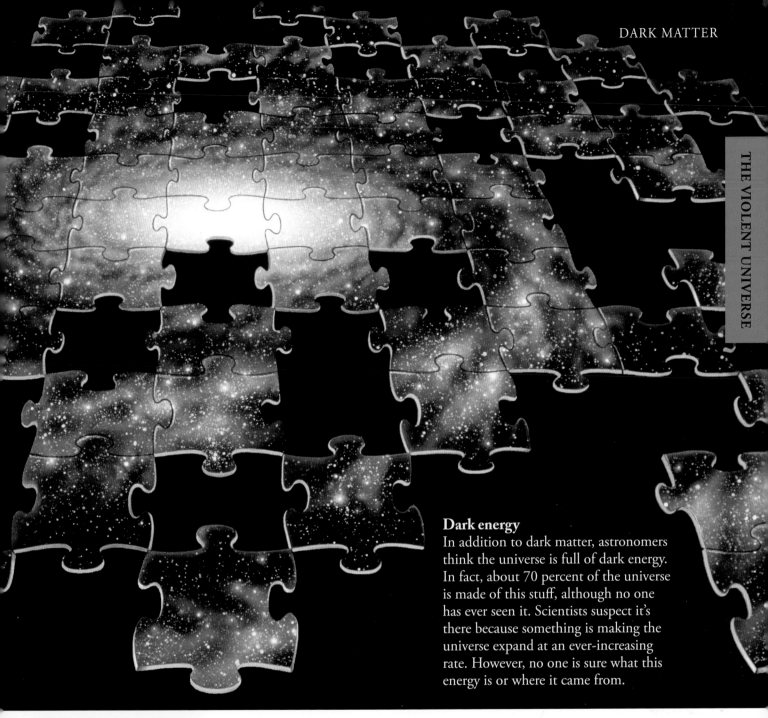

Dark energy

In addition to dark matter, astronomers think the universe is full of dark energy. In fact, about 70 percent of the universe is made of this stuff, although no one has ever seen it. Scientists suspect it's there because something is making the universe expand at an ever-increasing rate. However, no one is sure what this energy is or where it came from.

The Bullet Cluster

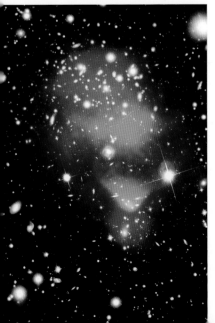

The Bullet Cluster was formed when two galaxy clusters collided, one tearing through the middle of the other like a bullet. The cluster's normal matter (which appears pink here) has been slowed down in the collision by a drag force. However, the dark matter has continued to move outward without slowing, creating a light-bending aura (shown in blue).

What's the matter?

The distant galaxy cluster Abell 383 contains large amounts of dark matter. Although dark matter is invisible, we can tell where it is from the way its gravity bends the light of distant galaxies.

◄ *The arclike shapes formed by the bending of light are evidence of the presence of dark matter in this cluster.*

LIFTOFF!

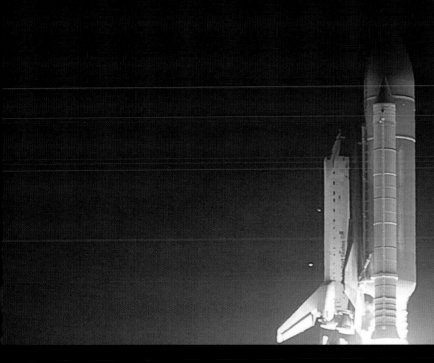

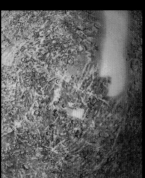

The first successful suborbital flight was made by a V-2 missile in 1942. But how do these huge, heavy machines take off, and what else have we sent into space?

How rockets *work*

A rocket is a launch vehicle used to carry astronauts or a payload (such as a satellite) from Earth into outer space. It must reach a speed of around 17,500 mph (28,000 kph) to overcome the pull of Earth's gravity and enter orbit. This is done by burning chemicals to create thrust.

Third stage delivers crew or payload to Earth orbit

Second stage takes over when the first stage is released

First stage includes the engines and fuel to launch the rocket

UNITED STATES

THRUST

GRAVITY

◀ IN PARTS *Each stage of a multistage rocket carries its own engines. When the fuel is used up, the stage is made to fall away.*

▲ NEWTON'S LAW *Isaac Newton's Third Law of Motion says, "To every action there is an equal and opposite reaction."*

LIFTING OFF

All objects on Earth are pulled down by gravity. So how does a huge, heavy rocket take off? As hot gases escape from a rocket's engines, they create a force called thrust that pushes the rocket in the other direction, against the pull of gravity. Isaac Newton explained that this works because every action (gases pushing down) has an equal and opposite reaction (rocket moving up).

ROCKET REGISTER

- **R-7 Semyorka** (Russian) Originally a missile, this was modified to launch Sputnik 1, the first artificial satellite.
- **Vostok** (Russian) In 1961, this was used for the first crewed space flight of cosmonaut Yuri Gagarin.
- **Saturn V** (American) The world's largest and most powerful rocket took the first people to the Moon in 1969.
- **Soyuz** (Russian) This family of rockets, first used in 1966, services the International Space Station.
- **Ariane** (European) Five types of Ariane have been used to launch satellites and probes into space.
- **Falcon** (American) Built by SpaceX, the various Falcon rockets have lower stages that can land safely and be reused.

ENGINES AND FUEL

■ **There are two types** of rocket engines: those that use solid propellant (fuel) and those that use liquid propellant. Many small rockets use solid propellant. Larger rockets may use a combination of solid fuel and liquid fuel in different stages.

■ **Boosters** are additional engines used to provide extra thrust for takeoff and are then jettisoned (thrown off).

■ **Solid fuel boosters** (shown below) are like fireworks: once they are lit, they cannot be shut down until all the propellant is used up.

Propellant   Casing   Burning surface

Liquid oxygen needed to burn the fuel

Nozzle

Liquid hydrogen

■ **Engines that use** liquid fuels (shown left) are much more complicated than solid-fueled boosters. This is because the fuel and propellant have to be stored in separate tanks, then brought together in a combustion chamber. This is where the fuel burns, creating hot exhaust gases.

Combustion chamber

Bring your own oxygen
To fly in space, rockets not only have to carry their own fuel, they also need to carry a source of oxygen, called an oxidizer. This is because chemicals (the fuel) need oxygen to burn, or combust. On Earth, oxygen is present in the air, but there is no oxygen in space for combustion. The combustion process generates hot gases that are directed out of nozzles at high speed, producing thrust.

▲ TESTING *The RS-68 rocket has liquid-fuel engines. Its exhaust gases are nearly transparent.*

Nozzles can be angled to change the direction of flight.

Booster

▲ REAR VIEW *Soyuz has four boosters around its core stage. The faster the hot gas escapes through the nozzles, the faster the rocket will fly.*

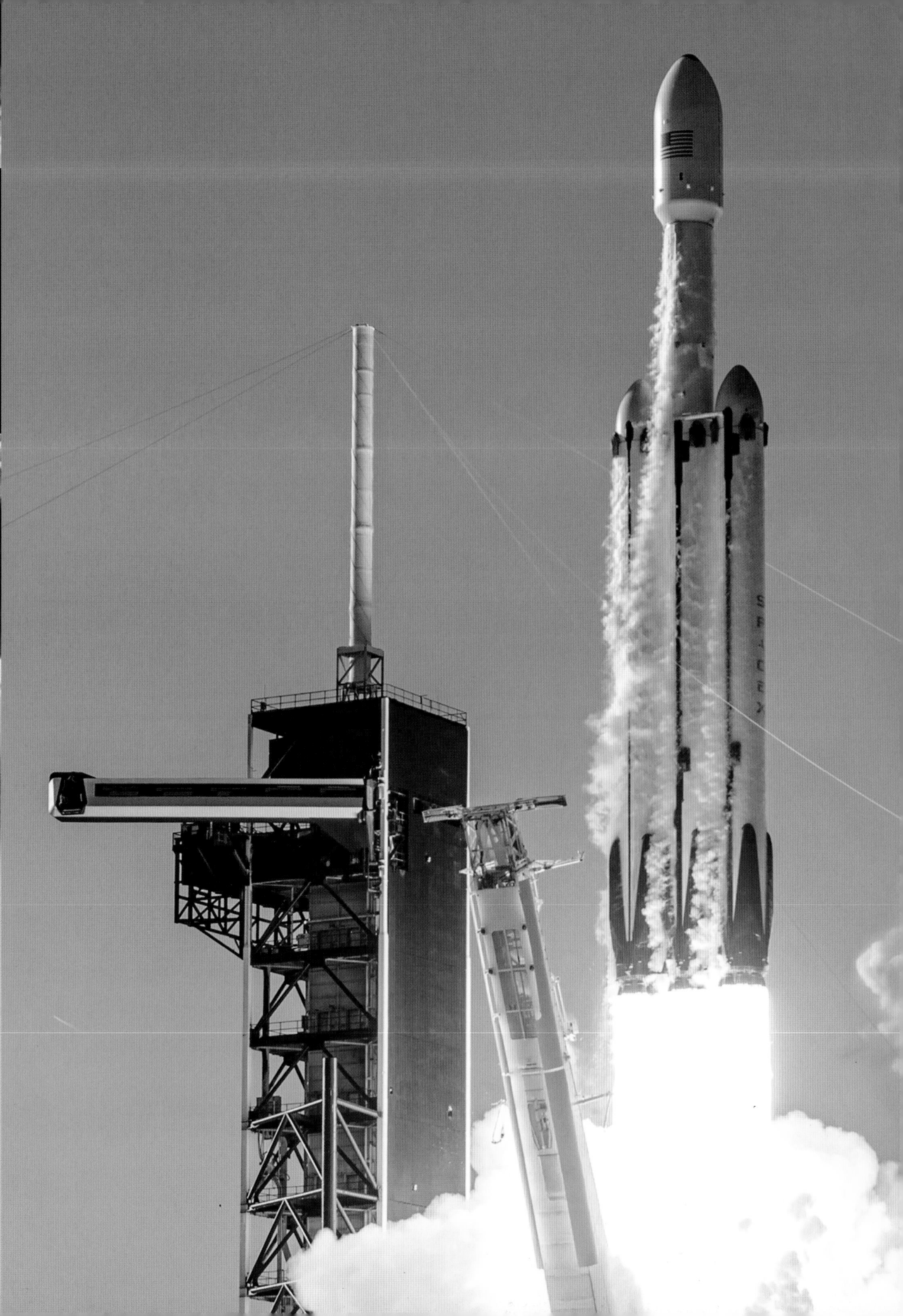

3, 2, 1 ...
... And the Falcon Heavy rocket blasts off into space to open up a new era of space exploration. Built by private company SpaceX, Falcon Heavy is the largest and most powerful rocket in operation today, combining three identical first stages from the company's reliable Falcon 9 launcher. It can put up to 63 tons (64 tonnes) of payload into low Earth orbit around 125 miles (200 km) above Earth.

The space *shuttle*

The space shuttle was the world's first reusable spacecraft. It took off like a rocket but landed back on Earth like a glider. The shuttle was launched for the first time in 1981 and flew on more than 130 missions before the program was retired in 2011. It carried a crew and cargo, and its missions included launching satellites and building space stations.

SHUTTLE COMPONENTS

The shuttle consisted of three main parts: a winged orbiter that carried the crew and the cargo, two white booster rockets, and a huge orange fuel tank. The fuel tank and the boosters were discarded, or jettisoned, during ascents—only the orbiter actually went into space. The fuel tank was the only part of the shuttle that could not be reused.

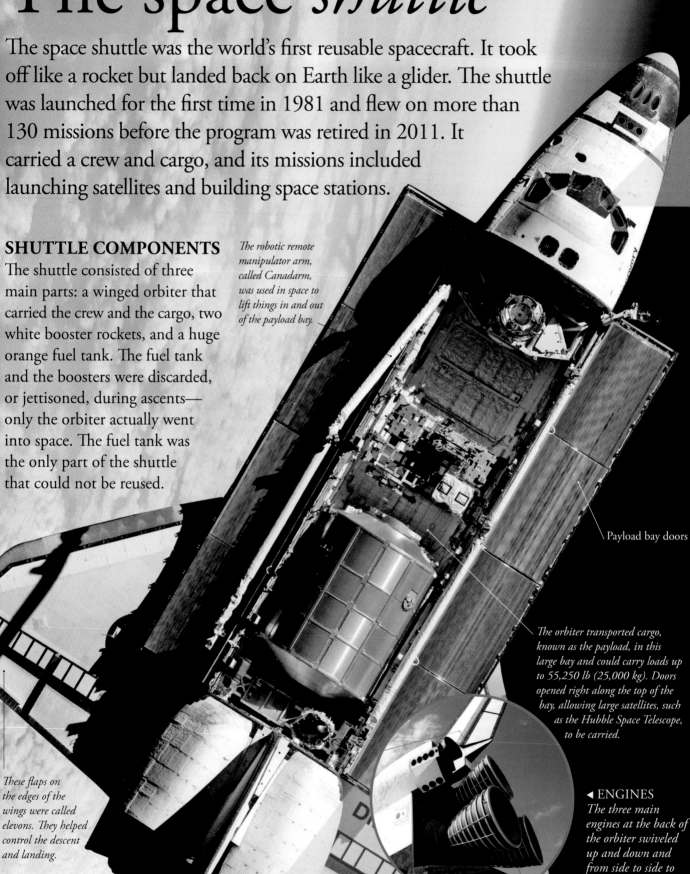

The robotic remote manipulator arm, called Canadarm, was used in space to lift things in and out of the payload bay.

Payload bay doors

The orbiter transported cargo, known as the payload, in this large bay and could carry loads up to 55,250 lb (25,000 kg). Doors opened right along the top of the bay, allowing large satellites, such as the Hubble Space Telescope, to be carried.

These flaps on the edges of the wings were called elevons. They helped control the descent and landing.

◄ ENGINES
The three main engines at the back of the orbiter swiveled up and down and from side to side to steer the shuttle.

The crew
On a typical mission, the shuttle carried five to seven crew members: a commander, a pilot, several scientists, and sometimes a flight engineer. They traveled in the crew compartment at the front of the orbiter, which contained the flight deck and their living quarters.

RETIREMENT
During its operating lifetime, the shuttle suffered two major disasters. In 1986, Challenger exploded moments after launch, and in 2003, Columbia broke up as it reentered Earth's atmosphere. Each accident cost the lives of seven astronauts, and the second led to a decision to retire the shuttle after the completion of the ISS space station in 2011.

LIFT-OFF!

Start of the journey
The space shuttle launched from the Kennedy Space Center in Florida. Lift-off was powered by the two booster rockets and the three main engines on the orbiter, which were fueled by liquid hydrogen and liquid oxygen from the fuel tank. About 2 minutes after lift-off, the booster rockets were jettisoned and fallen back to Earth. When the shuttle reached its orbit, the main engines were shut down, and the empty fuel tank was jettisoned and burned up in the atmosphere.

▲ CHALLENGER *exploded after a leak from one of its booster rockets.*

▼ HOME AGAIN *Atlantis deployed its drag chute when it landed in 2009.*

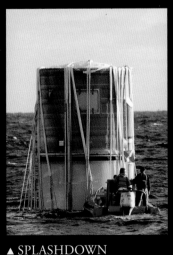

▲ SPLASHDOWN
The two booster rockets landed in the Atlantic Ocean, off the coast of Florida. They were recovered by ships so that they could be used again.

FAST FACTS
- The orbiter was 122 ft (37 m) long and had a wingspan of 78 ft (24 m).
- Only five space-worthy orbiters have ever been built: Columbia, Challenger, Discovery, Atlantis, and Endeavour.
- A typical mission lasted 12 to 16 days.
- The shuttle's main fuel tank held about 526,000 gallons (2 million liters) of fuel.
- During reentry, the outside of the orbiter heated up to more than 2,730°F (1,500°C).
- The shuttle could go from 0–17,000 mph (0–27,500 kph) in less than 8 minutes.

Touchdown
To leave orbit, the orbiter fired its thrusters and decelerated from hypersonic speed. It dropped down through Earth's atmosphere underside first, generating enormous heat through friction with the atmosphere. The shuttle landed on a long runway, usually at the Kennedy Space Center, using a drag chute to help it slow down.

Launch *centers*

The very first launch sites were located on military bases in the US and the USSR, and these have remained the main US and Russian launch centers ever since. Today, launch sites have been built or are under construction in many countries, including China, French Guiana, India, and South Korea.

▲ THE FIRST *launch pad built at Baikonur in the USSR was used to launch both Sputnik 1 and Yuri Gagarin, the first person in space (above), into orbit.*

AN IDEAL SITE

Rockets are not permitted to take off over highly populated areas, so launch sites are always located in remote places. A site near the sea, such as Cape Canaveral on the Florida coast, works well. Rockets launch eastward, over the Atlantic Ocean, and any jettisoned stages fall into the water.

▼ THIS ROCKET, *shown in the vehicle assembly building at the Kennedy Space Center, was the first Saturn V to be launched. It was used on the Apollo 4 mission.*

Cape Canaveral (US)

This launch center started life as a missile test center located on the site of an old air base. The first rocket was launched there in 1950. Since 1958, the site has been the main center for US launches and the only one for crewed missions. Launch Complex 39, located on an island to the north of Cape Canaveral, was added in the 1960s for Saturn V launches. This area is known as the Kennedy Space Center. In total, more than 500 rockets have been launched from the Cape.

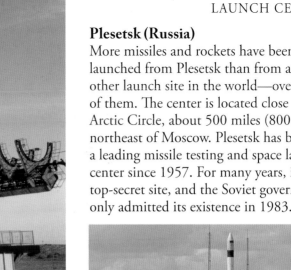

Plesetsk (Russia)

More missiles and rockets have been launched from Plesetsk than from any other launch site in the world—over 1,500 of them. The center is located close to the Arctic Circle, about 500 miles (800 km) northeast of Moscow. Plesetsk has been a leading missile testing and space launch center since 1957. For many years, it was a top-secret site, and the Soviet government only admitted its existence in 1983.

▲ THE PLESETSK *launch site is situated in an area of forest and lakes. About 40,000 service personnel and their families live in the nearby town of Mirnyy.*

Baikonur (Russia)

All Russian crewed flights and planetary missions are launched from Baikonur, a center situated on the flat, deserted plains of neighboring Kazakhstan. The Baikonur "cosmodrome" includes dozens of launch pads, nine tracking stations, and a 930 mile (1,500 km) long rocket test range. Missile and rocket tests started there in 1955.

▶ ARIANE 5 *rockets are launched from the site at Kourou. They carry payloads for the European Space Agency.*

Kourou (European Space Agency)

The location of this launch site in French Guiana is one of the best in the world. It is near the equator, which gives the maximum energy boost from the Earth's rotation for launches into equatorial orbits, and weather conditions are favorable throughout the year. The site has been used as the main European spaceport since July 1966. A new pad has recently been built for use by the Russian Soyuz launcher.

Jiuquan (China)

This launch center is situated in the Gobi Desert, 1,000 miles (1,600 km) west of Beijing, and was first used in 1960. In 1970, a Long March-1 rocket launched the Mao-1 satellite from Jiuquan, making China the fifth nation to launch an artificial satellite into orbit. Today, Jiuquan is the launch site for China's crewed Shenzhou spacecraft, but it is limited to southeastern launches to avoid overflying Russia and Mongolia.

The Odyssey (Sea Launch company)

The most unusual launch site is the Odyssey platform, which launches rockets from the middle of the Pacific Ocean. A satellite is prepared onshore in California, attached to a Zenith rocket, then transferred to the Odyssey platform. The platform sails to a site near the equator, a journey of 11 to 12 days, then the rocket is launched.

Launching *Ariane 5*

Launched from the Kourou spaceport in French Guiana, the Ariane 5 rocket is capable of lifting two satellites weighing almost 10 tons (9 tonnes) into orbit. The rocket and its satellites are assembled and prepared for launch in special facilities at the Ariane launch complex.

THE LAUNCH COMPLEX

The ELA-3 launch complex was built in the 1990s especially for the European Ariane 5 rocket. Between eight and 10 rockets can be launched there each year, and each launch campaign lasts 20 days. The control center is located in a protected enclosure, designed to withstand the impact of any falling rocket pieces and has two independent launch control rooms.

▼ TECHNICIANS *load the Philae lander onto the Rosetta probe ready for its journey to comet Churyumov-Gerasimenko (*👁 *p. 157).*

▼ A SOLID *rocket booster arrives for integration with an Ariane 5 rocket at the assembly building.*

Rocket stages

The 190 ft (58 m) high launcher integration building is where the stages of the Ariane 5 rocket are joined together. The rocket's core stage is positioned on a mobile launch table and the two solid boosters are attached on either side. The core stage is equipped with one of three available upper stages. The launch table and the rocket are then transferred to the final assembly building.

▲ THE MAIN STAGE, *which will contain the liquid propellant, is hoisted into position and the nozzle is attached.*

Preparing the payload

Satellites are prepared for launch in the vast payload processing building. It is so big that several satellites can be handled at once. The building also has two areas for hazardous activities, such as loading the highly inflammable propellant (fuel). The finished payload, now ready for launch, is then taken to the assembly building to be attached to the rocket.

◄ AN ARIANE 5 *rocket consists of a central core stage, two solid boosters, and an upper stage. It is almost 170 ft (52 m) high.*

► WATER TOWER *This supplies the water that is showered into the flame trenches and around the launch table. It holds about 400,000 gallons (1.5 million liters) of water.*

LIFTOFF!

Final assembly

Inside the final assembly building, the satellite is installed on top of the rocket. It is covered with a shell, known as the payload fairing, which protects the satellite during the launch. Then the rocket's upper stage and the attitude control system are fueled. About 12 hours before the launch, the mobile launch table and the completed rocket are rolled out to the launch zone.

► THE ROCKET *is slowly moved out on a crawler tractor.*

▲ THE PAYLOAD *is hoisted by a special mobile crane and placed on top of the rocket.*

Launch zone

This area is where the most dangerous operations take place, so it is located 1.7 miles (2.8 km) from the other buildings. The rocket's core stage is filled with liquid hydrogen and liquid oxygen propellant, then the main engine and solid booster stages are ignited and the rocket lifts off. The launch zone has a concrete foundation with three flame trenches that catch the rocket's exhaust. During liftoff, the area is showered with water to reduce the effects of noise and heat.

Artificial *satellites*

In astronomy, a satellite is a body that orbits a planet. There are natural satellites, such as moons, and artificial (human-made) satellites, such as communications satellites and space stations. The first artificial satellite was very simple, but modern ones are much more complicated.

Four aerials on Sputnik transmitted radio signals.

IT'S GOOD TO TALK

Many artificial satellites are designed for communication—sending data such as TV broadcasts, mobile phone signals, pictures of clouds and land use, and scientific information. The owners of a satellite also need to be able to keep track of it. This is mainly done using dish-shaped antennas on the ground and on the satellite.

Sputnik 1

Launched on October 4, 1957, the Russian satellite Sputnik 1 was the first artificial satellite to successfully be placed in orbit around Earth. The 23 in (58 cm) diameter aluminum ball carried four radio antennas up to 10 ft (3 m) long. Sputnik's beeping signals continued for 21 days, but it survived in orbit for 92 days before burning up during reentry on January 4, 1958.

◀ LASER LOCATOR
The precise orbits of some satellites are checked by bouncing laser pulses off the satellite.

I'VE GOT THE POWER

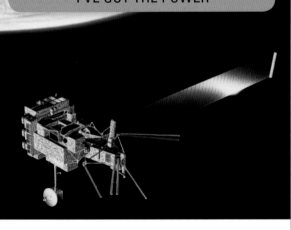

■ Satellites have to power themselves. This is usually done by using large solar arrays ("wings") crammed with light-sensitive solar cells. The arrays are many feet long and have to be folded during launch.
■ The solar cells can provide several kilowatts of power, although they become less efficient as they get older.
■ The arrays can be turned so that they always collect as much sunlight as possible. When the satellite goes into shadow, it gets its power from rechargeable batteries.

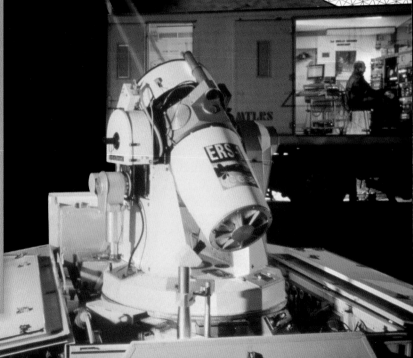

DON'T MISS THE BUS

Most commercial satellites are built on the same basic model, designed to be as strong and light as possible. A platform called a bus contains all the main systems, including the batteries, computer, and thrusters. Attached to the bus are antennas, solar arrays, and payload instruments (such as cameras, telescopes, and communications equipment) that the satellite uses to do its job.

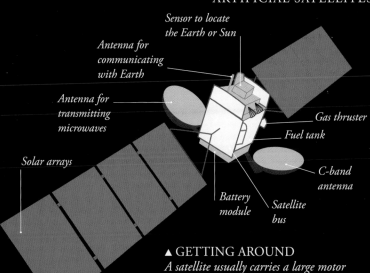

Sensor to locate the Earth or Sun

Antenna for communicating with Earth

Antenna for transmitting microwaves

Solar arrays

Gas thruster

Fuel tank

C-band antenna

Battery module

Satellite bus

LIFTOFF!

▲ GETTING AROUND
A satellite usually carries a large motor and thrusters to move the satellite into the correct position once it separates from its launch vehicle.

Super solar satellite
Vanguard 1 holds the record for being the oldest human-made object in space. Launched in 1958, it was the fourth artificial satellite to successfully reach orbit and the first to be powered by solar panels. Communication with Vanguard stopped in 1964, but the satellite still circles the Earth among a cloud of space debris.

A satellite can be affected by many things. Small meteorite impacts, the solar wind, solar radiation, and minor changes in gravity can all alter its position or even cause damage.

◀ POINT IT RIGHT
Many satellites need to point in the right direction to line up their antennas and communicate with Earth. Getting the correct position, or "attitude," can be a tricky job!

Hot and cold
The side of a satellite facing the Sun gets very hot, while the shaded side becomes very cold. This causes problems because most satellite equipment is sensitive to extreme heat or cold. Ways of protecting equipment include using layered insulating blankets that look like foil to keep heat in and adding radiators to release heat from electrical equipment.

NASA's Lunar Reconnaissance Orbiter (LRO) is a robotic spacecraft sent to study the Moon's surface from an orbit 31 miles (50 km) away.

Satellites *in orbit*

Thousands of satellites have been sent into space since Sputnik 1 in 1957. There are many different types and sizes and many different uses. Most are placed in low Earth orbit between 125 and 1,250 miles (200 and 2,000 km) above Earth. These take about 2 hours to orbit Earth once.

WEATHER WATCHER

Some weather satellites, such as the European Space Agency's Meteosats, are in geostationary orbit—they stay above the same place on Earth. Orbiting 22,240 miles (35,800 km) above Earth, they take 24 hours to go once around the planet. With cameras facing one side of Earth, they can study changing weather.

Weather forecasting
Satellites, especially those in low polar orbits, can take amazingly detailed images of weather. They are used to forecast the weather—but we don't always get it right! The image below, taken by NASA's Terra satellite, is of tropical cyclone Gonu passing over the Gulf of Oman. The storm was predicted to travel inland, but it didn't.

▲ STAYING POWER *This Meteosat stays above West Africa, on the equator. As Earth turns, the satellite follows.*

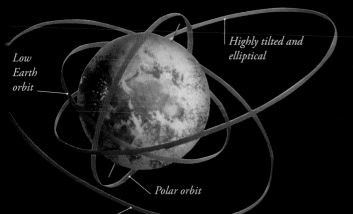

Low Earth orbit

Highly tilted and elliptical

Polar orbit

Geostationary orbit

SAT NAV FAMILIES

- There are several families of satellites that provide navigation information. The best known, and most widely used, is the American Global Positioning System (GPS).
- GPS has over 30 satellites in 6 orbits that criss-cross 15,000 miles (24,000 km) above Earth. There are nearly always three or four satellites above the local horizon at any one time.
- Russia's Glonass system is similar to GPS.
- Europe's Galileo Satellite System is due to be completed in 2020.

Types of orbit

Different orbits are used for different missions. Many communications and weather satellites stay above the equator, either in a near orbit called low Earth orbit, or much farther out, in a geostationary orbit. Satellites can survey the entire planet in great detail from low polar orbits as Earth spins beneath them. Earth observation satellites and astronomical observatories can be found in highly elliptical (oval-shaped), tilted orbits.

Satellite navigation

Many cars, trucks, and aircraft are fitted with satellite navigation equipment (sat-nav) that acts as an electronic map and route finder. They work by picking up signals from four satellites at the same time, which locates your precise position on the planet.

▶ GALILEO GUIDE *A European sat-nav system called Galileo is currently being launched. It will have 30 satellites in three inclined (tilted), circular orbits.*

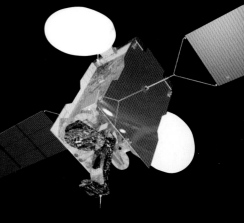

Observing Earth

Many satellites are used to study Earth's surface. From their images, we can learn about many subjects, including changing land use, ocean currents, and air pollution. By taking pictures of the same place from different angles, they can produce 3D images. Some satellites can see objects smaller than 20 in (50 cm) across and may even be able to read headlines on a newspaper. Radar satellites can see the ground even at night or when an area is covered by clouds.

Telecommunications satellites

Radio, TV, and telephone communications have been transformed by satellite technology. The first live TV signals were relayed from America to Britain in 1962. Today, satellites can transmit hundreds of digital TV channels to rooftop dishes. We can watch live events and sports tournaments from around the world, and satellite phones make it possible to call someone in the middle of a desert or on top of a mountain.

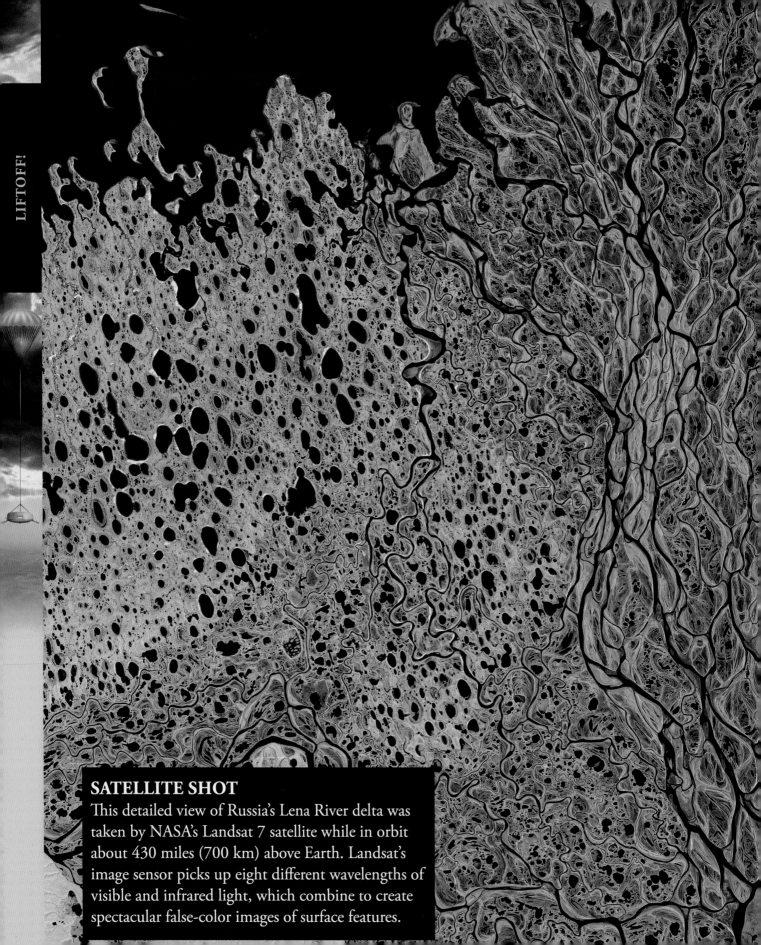

SATELLITE SHOT

This detailed view of Russia's Lena River delta was taken by NASA's Landsat 7 satellite while in orbit about 430 miles (700 km) above Earth. Landsat's image sensor picks up eight different wavelengths of visible and infrared light, which combine to create spectacular false-color images of surface features.

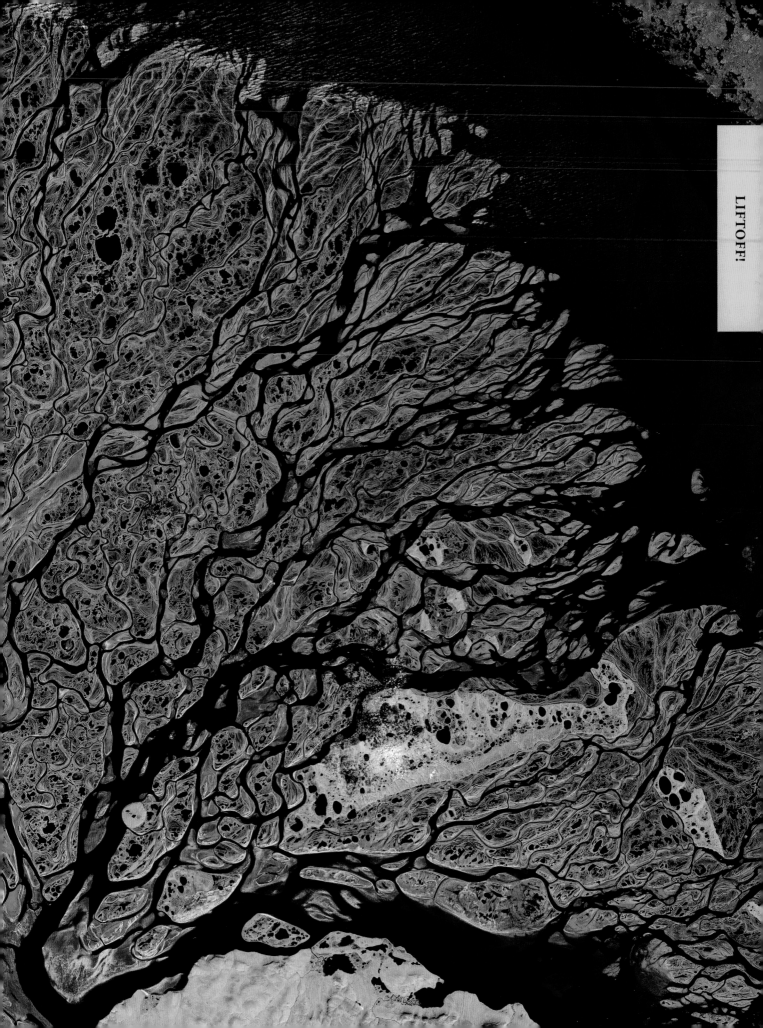

LUNA 3's CAMERA *took 29 photos over 40 minutes, imaging 70 percent of the previously unseen far side.*

Space *probes*

In the 1950s and 1960s, the Soviet Union and the US sent the first uncrewed spacecraft, or probes, to explore the Moon, Venus, and Mars. Since then, probes have visited the Sun; all the other planets in our solar system; and many moons, asteroids, and comets.

The far side of the Moon

In January 1959, the Soviet probe Luna 1 became the first spacecraft to fly past the Moon. This was followed in October 1959 by Luna 3, which sent back the very first images of the far side of the Moon. Luna 3 was launched into an elliptical (oval-shaped) Earth orbit that enabled it to swing behind the Moon, just 3,850 miles (6,200 km) above its surface. The onboard camera took photographs of the far side, which revealed that it has very few "seas."

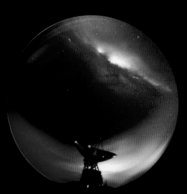

▲ THE MILKY WAY *in the night sky over a spacecraft-tracking antenna.*

TRACKING PROBES

Probes transmit images and other data back to Earth in the form of high-frequency (short-wavelength) radio waves. This information is picked up by dish-shaped tracking antennas at satellite ground stations.

Mars

Phobos, one of Mars's two moons.

FIRST PLANET ORBITER
The US probe Mariner 9 was launched in May 1971, on a mission to orbit the planet Mars. It sent back remarkable images of huge volcanoes, a vast canyon system, dry river beds, and close-up pictures of its two moons.

First mission to Venus

The Mariner series were the first US probes to be sent to other planets. Mariner 2 was launched successfully in July 1962 and flew past the planet Venus at a distance of 21,648 miles (34,838 km) on December 14, 1962. The probe scanned the planet for 42 minutes as it passed, revealing that Venus has cool clouds and a very hot surface, with temperatures of at least 800°F (425°C).

▲ MARINER 2 *The spacecraft had a conical frame of magnesium and aluminum, with two solar panels and a dish antenna.*

Journey to Jupiter

Pioneer 10 was launched in March 1972, and became the first spacecraft to travel through the asteroid belt (between July 1972 and February 1973) and the first to reach the planet Jupiter. The probe sent back close-up images of Jupiter, then continued on its journey out of the solar system, crossing Neptune's orbit in May 1983. The last signal was received from the probe in 2003. Pioneer 10 is heading for the star Aldebaran, in the constellation Taurus, but it will take more than 2 million years to get there!

Mission to Mercury

In 1974, Mariner 10 became the first spacecraft to visit the planet Mercury. It was also the first to use another planet's gravity to alter its course when it flew past Venus on February 5, 1974. The first Mercury flyby took place on March 29, 1974, with two more over the following months. The probe sent back 12,000 pictures of Mercury, which showed a heavily cratered world, much like our Moon.

▶ MISSING AREA *Mariner 10 was unable to see this part of the planet's surface.*

First planetary balloons

The two Soviet probes Vega 1 and 2 were launched in December 1984 on a mission to fly past Venus. They released two landers and two instrument packs, attached to Teflon-coated balloons, into the planet's atmosphere. Both balloons survived for about 46 hours and sent back data on the clouds and winds, while the landers explored the lower atmosphere and surface rocks.

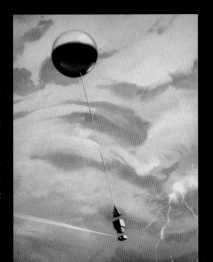

FACT FILE

■ When it was launched, Pioneer 10 was the fastest spacecraft to ever fly. It left Earth at a record-breaking 32,107 mph (51,670 kph).

■ For many years, Pioneer 10 was the most remote human-made object in the solar system, but on February 17, 1998, it was overtaken by the probe Voyager 1.

■ Vega 1 and 2 flew on from Venus to fly past Halley's Comet in March 1986.

Vega probes were powered by solar panels and carried an antenna dish, cameras, and an infrared sounder.

Space *debris*

More than 2,000 satellites are currently in operation, most of them in orbit around Earth. However, these satellites are flying through an ever-increasing sea of space debris. This debris field includes objects ranging from the size of a car to tiny specks of dust and paint.

▼ OUTER RING *This consists mainly of debris from telecommunications satellites.*

▶ LOW EARTH ORBIT
About 70 percent of the debris is in low Earth orbit, which extends to 1,250 miles (2,000 km) above Earth's surface. The objects are most closely spaced at high latitudes above the polar regions.

WHERE IS THE DEBRIS?

At present, there are more than 20,000 pieces of debris more than 4 in (10 cm) across and millions of smaller pieces orbiting our planet. The majority of them are in low Earth orbit, but there is a second ring of debris at an altitude of about 22,000 miles (36,000 km), an orbit used mainly by communications satellites. This ring is rapidly filling up, so most elderly satellites are now boosted into a higher "graveyard" orbit before they are shut down.

Falling to Earth
Pieces of debris that fall into Earth's atmosphere normally burn up, like man-made shooting stars. But occasionally an object reaches the ground almost intact. This propellant tank from a Delta 2 rocket landed in Texas in 1997.

EXPLOSIONS

So far, there have been more than 200 explosions in space, and more are very likely. Explosions are usually caused by uncontrolled events, such as pressure build-up in a rocket's fuel tanks, battery explosions, or the fuel igniting. Each explosion creates thousands of small fragments of debris.

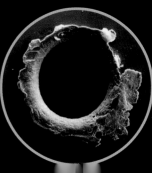

Collisions

The first recorded collision between two large objects took place in 1996, when the French Cerise satellite was hit by a fragment from an Ariane rocket. In 2009, two satellites, Cosmos 2251 and Iridium 33, actually collided (above). The resulting explosion created a massive cloud of debris—perhaps 100,000 pieces of junk.

The Cerise satellite collided with a piece of debris from an Ariane rocket, which tore off a piece from the boom, leaving the satellite severely damaged.

Astronauts at risk

Crewed spacecraft, such as the Soyuz, fly in low Earth orbit, where debris is most common. The US military tracks big pieces of debris and issues a warning if a close encounter is likely. Spacecraft can then avoid the danger, but hits from small debris are unavoidable. In 54 shuttle missions up to 2005, space junk and small meteorites hit the windows 1,634 times.

▲ PIECE OF DEBRIS
This fragment measures about 2 in (5 cm)—big enough to cause major damage to a spacecraft.

◄ WINDOW DAMAGE
Shuttle windows often have to be replaced because of chips in the glass caused by debris.

► DEBRIS HOLE
This is a hole in a panel on SolarMax, a satellite monitoring solar flares.

FAST FACTS

■ Even tiny pieces of debris can cause a lot of damage, because they are traveling at speeds around 17,000 mph (27,000 kph). The high speed turns a fleck of paint into the equivalent of a rifle bullet.

■ The International Space Station is fitted with special shields to protect its skin from debris impacts. It can also be moved out of harm's way if a particularly large piece of debris poses a threat.

■ Optical telescopes and radars are used to track large pieces of debris from the ground.

■ The amount of human-made debris in space is expected to grow in the future, even if there are no more explosions. This is because collisions between pieces of debris will create dozens, or even hundreds, of smaller fragments.

Space *nations*

For many years, space exploration was dominated by two countries—Russia and the United States. However, over time, more countries built their own satellites and launch rockets. Today, a new generation of space powers—including China, India, Brazil, South Korea, and Israel—is prepared to spend large sums on developing its space industry.

FAST FACTS

■ It only takes about 10–30 minutes for a rocket to put a satellite in orbit.
■ Chinese-Brazilian satellites can get very detailed photographs of cities from 435 miles (700 km) away.
■ The US Space Surveillance Network tracks objects in space; at present, there are 2,000 satellites operating above Earth.
■ Satellites that do not appear to move through the sky are, in fact, orbiting at the same speed as Earth rotates.

ROCKET FLEETS

To get their satellites into orbit, many smaller countries book a ride on a European, Russian, or Japanese rocket, but India and China now have launch sites and reliable rocket fleets that can be used instead. India's Polar Satellite Launch Vehicle (PSLV) has launched more than 200 satellites, including 104 in one go in 2017. Israel has a small launcher, while Brazil, Iran, and North and South Korea are developing their own rockets and launch sites.

TAKE A LOOK: OVER THE MOON

In 2009, the Indian lunar orbiter Chandrayaan-1, with NASA equipment on board, sent back data that indicated that water was present in the Moon rock. The discovery was backed up by previous data collected by two US spacecraft, Cassini and Deep Impact.

▲ THIS INFRARED *image of a crater on the far side of the Moon looks quite dry and dusty.*

▲ HOWEVER, *when the crater is seen in false color, there is widespread evidence of water in the rocks and soil.*

CREWED MISSIONS

So far, the only new country to put a human in space is China. In 2003, they sent a single astronaut (or "taikonaut" in Chinese), Yang Liwei, into orbit. The second mission in 2005 carried two astronauts. On the third mission in 2008, Zhai Zhigang became the first Chinese person to space walk. In 2012, China's first woman in space, Liu Yang, joined a mission to the first Chinese space station.

▲ THE THREE-MAN CREW *on China's third crewed space mission Shenzhou-7 were treated like celebrities both before and after their trip into space.*

INTO ORBIT

Satellites are used for many different things. Countries such as India, Brazil, China, and South Korea have been sending up rockets carrying survey satellites that can help them monitor the weather and pollution, look for minerals and resources, or check on farming or urban areas. Others carry telecommunications or global positioning equipment.

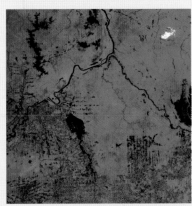

▲ ENVIRONMENTAL *monitoring by the joint China-Brazil Earth Resources Satellite (CERBS-1) has located areas of deforestation (shown here in pink) in the Amazon rainforest.*

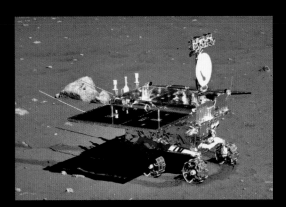

China's Chang'e-3

In 2013, the Chang'e-3 probe made a soft landing on the Moon and deployed a small robot rover called Yutu ("Jade Rabbit"). Yutu studied the composition of the Moon's soil up to 100 ft (30 m) deep.

▼ THE JAPANESE *Aerospace Exploration Agency (JAXA) is a major player in space exploration today. It uses its own rockets to launch its satellites and spacecraft. SELENE was launched by its H-IIA rocket.*

HIGH-DEFINITION MOON

In September 2007, Japan launched its SELENE orbiter, nicknamed Kaguya after a legendary princess. It was the biggest lunar mission since Apollo. The aim of the mission was to investigate the Moon's origin and evolution, but Kaguya also carried a high-definition video camera that filmed a sensational movie of Earth rising over the lunar horizon.

WATCH THIS SPACE

The JAXA spacecraft also mapped the Moon's rugged terrain in 3D and studied its magnetic field. The mission was a great success, and in 2009, after 1 year and 8 months, the orbiter had a planned crash-landing onto the Moon.

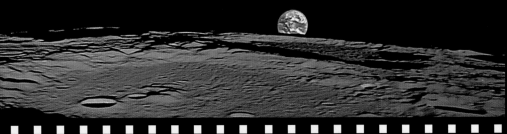

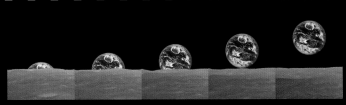

◄ THE AMAZING *video of Earth rising was loaded onto YouTube and has been viewed by over a million people.*

Super spacecraft

Traveling through space can take a very long time. Robotic spacecraft have flown huge distances to explore most of the solar system, but the difficulties of people traveling to Mars and beyond have yet to be solved. However, many ideas are being tested to speed up space travel and save on fuel. Could these lead to crews exploring distant worlds in the not-too-distant future?

POWERED BY ELECTRICITY

Traditional rocket engines burn large amounts of fuel. This makes the vehicles very large and heavy and very expensive to fly. Electric propulsion—also known as an ion drive—is much lighter and more efficient. It works by firing a stream of electrically charged particles (ions) into space. The ions pass through an electrically charged grid, which makes them move very fast. The thrust is weak, but over time, it can propel the spacecraft to very high speeds.

The European Space Agency's SMART-1 lunar probe is powered by an ion drive.

SMART-1 probe

Moon

Earth

This is where SMART-1 escaped Earth's gravity and was pulled into orbiting the Moon.

SMART MOVES

Launched in 2003, SMART-1 was the first European spacecraft to use the Moon's gravity to pull it into orbit. First, it spiraled around Earth on an ever-enlarging orbit, firing its ion drive to turn the natural circular path into an ellipse (oval). When it was far enough away to escape Earth's gravity, it was pulled into a new orbit by the Moon.

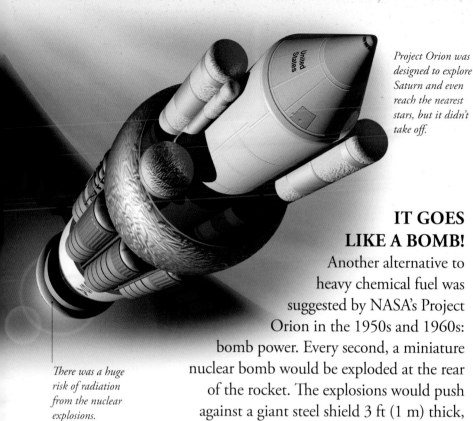

Project Orion was designed to explore Saturn and even reach the nearest stars, but it didn't take off.

There was a huge risk of radiation from the nuclear explosions.

IT GOES LIKE A BOMB!

Another alternative to heavy chemical fuel was suggested by NASA's Project Orion in the 1950s and 1960s: bomb power. Every second, a miniature nuclear bomb would be exploded at the rear of the rocket. The explosions would push against a giant steel shield 3 ft (1 m) thick, propelling the rocket up and into space.

AEROBRAKING

Spacecraft use a lot of fuel as they brake into orbit around the Moon and planets. However, if the planet has an atmosphere, it is possible to slow down without using a rocket engine. This is done by dipping in and out of the upper atmosphere—a process known as aerobraking. Each time the spacecraft enters the atmosphere, it is slowed a little by friction. This technique can also be used to change its orbit.

Mars Reconnaissance Orbiter aerobraking.

PROJECT DAEDALUS

In the 1970s, the British Interplanetary Society's Project Daedalus described a two-stage, uncrewed craft that would be built in Earth's orbit. Its engines would use nuclear fusion—the same power source as the Sun—to fire high-speed jets of gas into space. Nearly all of its 60,000 ton (54,000 tonne) weight would be fuel. While it would be fast enough to reach Barnard's Star (almost 6 light-years away) within 50 years, it would need as much fuel to slow down as to accelerate, so it would just speed past the star and keep on going.

The IKAROS sail is 65 ft (20 m) across but only 0.0003 inch (0.0075 mm) thick.

◀ *The project name for the first Japanese solar sail mission is IKAROS (Interplanetary Kite-craft Accelerated by Radiation Of the Sun).*

SOLAR SAILS

Sailing ships have been used on Earth for thousands of years, but soon there may be sails in space. The idea behind solar sails is that sunlight pushes down on solid surfaces, so if enough light was bounced off a large, lightweight sail, it could push a spacecraft through space. The thrust would be small but continuous, and over time, the spacecraft could reach high speeds.

HUMANS IN SPACE

Living in space is not easy. From preflight training to building a space station in orbit, there is a lot of work for astronauts to do—in zero gravity, a long way from home.

Space *pioneers*

Since the 19th century, many people—
and animals—have taken part in
mankind's efforts to develop spacecraft
and explore outer space. Here are a select
few whose contributions changed the
course of history.

Space animals
Animals were sent
into space in the 1940s
and 1950s to see how
weightlessness affected
them. Two monkeys, Able
and Miss Baker, were launched
300 miles (483 km) above the
Earth in 1959 and experienced
weightlessness for 9 minutes
before returning safely to Earth.

Konstantin Tsiolkovsky

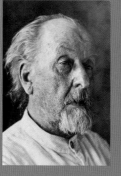

"The Earth is
the cradle of
mankind—one
cannot remain
in the cradle
forever." So
said Konstantin
Tsiolkovsky, a
Russian rocket scientist and pioneer
of human space travel. He first
became interested in spaceflight
in 1874, when he was only 17. He
went on to write about his ideas
for multistage rockets, liquid-fueled
propulsion, pressurized spacesuits,
and orbital space stations. These
theories were used to develop space
exploration after his death in 1935.

*Verne's spacecraft was
fired from a huge cannon
called Columbiad. NASA
used the name Columbia
for the command module
that took man to the
Moon in 1969.*

Jules Verne
Jules Verne was a
science-fiction writer in the
19th century. His story *From the
Earth to the Moon* and its sequel inspired
many space pioneers, including Konstantin
Tsiolkovsky, Robert Goddard, and Wernher
von Braun.

*Goddard worked on his
own, conducting many
practical experiments with
his rockets in the 1920s.*

Robert Goddard
People thought the American
physicist Robert Goddard was crazy
when he first began developing his
ideas on rocket propulsion and
spaceflight. His first liquid-fueled
rocket was successfully launched
at his aunt Effie's farm in 1926.
His 10 ft (3 m) rocket went 41 ft
(12.5 m) high, traveled 184 ft (56 m),
and flew for only 2.4 seconds. Now,
Goddard is recognized as one of
the fathers of modern rocketry.

> I could have gone on flying through space forever.

Gagarin had to parachute from the capsule before it landed—although this was kept a closely guarded secret for many years.

Yuri Gagarin—first person in space

A keen jet-fighter pilot, Yuri Gagarin became a cosmonaut candidate in 1959. On April 12, 1961, his Vostok spacecraft was launched into orbit 203 miles (327 km) above Earth. Traveling at 17,500 miles an hour (28,000 kph), his single orbit around the Earth only lasted 108 minutes, but it caused a sensation and made him world famous.

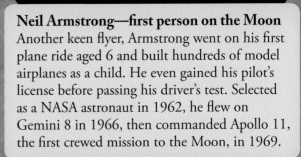

▼ ONLY A DOZEN *men have set foot on the Moon, with Neil Armstrong leading the way on July 20, 1969.*

Neil Armstrong—first person on the Moon

Another keen flyer, Armstrong went on his first plane ride aged 6 and built hundreds of model airplanes as a child. He even gained his pilot's license before passing his driver's test. Selected as a NASA astronaut in 1962, he flew on Gemini 8 in 1966, then commanded Apollo 11, the first crewed mission to the Moon, in 1969.

Dr. von Braun standing by the engines of the Saturn V rocket.

Wernher von Braun

Originally, von Braun lived in Germany, where he developed the V-2 rockets used as weapons during World War II. After the war, he worked in the US on the Saturn V rockets, which helped the Americans win the race to the Moon. The Saturn V was famous for being the only rocket that worked every time without blowing up!

Sergei Korolev

An enthusiastic experimenter with rockets, Sergei Korolev attracted the attention of the Russian military in the 1930s and became the mastermind behind the development of the Russian space program, including the world's first artificial satellite, Sputnik. The Russians, however, kept his identity a secret, and he was only known as "Chief Designer" until after his death in 1966.

Becoming an *astronaut*

Becoming an astronaut is far from easy. Thousands of people apply, but only a few are chosen. Those selected have to undergo months of study and training before they can fly in space. Some astronauts say that the training is harder than the actual mission.

ASTRONAUT NEEDED!

Do you have the necessary qualifications to be an astronaut?
- *University degree:* engineering, science, or mathematics
- *Physically fit and healthy*
- *Good people skills*
- *Able to work in a team*

To be a mission specialist, you also need:
- *An advanced degree*
- *Professional experience:* engineering or space-related occupation

To be a pilot, you also need:
- *Flying experience:* Years of flying experience (usually in high-performance military jets)

THE CHOSEN FEW

In the early years of the space age, the only people chosen as astronauts were young military pilots with the highest levels of physical and mental toughness. Today, astronauts experience much lower stresses during lift-off and reentry, but they still have to pass an intensive physical examination.

WHAT A STAR!

US senator John Glenn has broken two space records: in 1962, on the Friendship 7 mission, he became the first American to orbit Earth, and in 1998, aged 77, he became the oldest person to go into space when he went up on the space shuttle.

Have you got what it takes?

Each country has its own training schedule, but they usually follow the same guidelines. Training lasts for approximately 2 years and typically covers about 230 subjects, including scuba diving, space engineering, language skills (English and Russian), space-walk training, and how to live and work in a zero-gravity environment—some 1,600 hours of instruction in all. It's hard work, and you have to be extremely dedicated, but what a reward at the end!

ASTRONAUT TRAINING **LOG BOOK:**

5, 4, 3, 2, 1 … lift-off!

FEBRUARY

We get to learn how to fly a spacecraft in flight simulators: from lift-off, to landing, to reentering Earth's atmosphere … again and again and again. Practice makes perfect!

MARCH

Have to train in the gym regularly to keep fit—being an astronaut is a very physical job.

APRIL

I love learning how to fly T-38 high-performance jets. Had to practice escaping from one sinking underwater last week. Learning how to use the ejector seat and a parachute, too.

Went swimming in a tank with a full-size replica of a spacecraft! Underwater, the normal pull of gravity isn't as strong, and we got to know every inch of the craft, inside and out. We also rehearsed space walks.

Today went swimmingly!

JULY

To get us used to weightlessness, we had to travel in a special padded plane. The pilot gave us a rollercoaster ride— hard not to feel sick, but fun playing at being Superman!

This plane is known as the "vomit comet."

Survival training in the jungle.

OCTOBER

We had to learn survival techniques in case we crash-landed in the jungle or somewhere cold after reentry. We are given medical training, too. We need to work as a team.

NOVEMBER

Winter training! Cold and hungry.

DECEMBER

We've been given our missions and are busy studying in the classroom now.

WHAT A CHORE!

Michael Lopez-Alegria, astronaut at Johnson Space Center in the US, said that while training, learning how to brush his teeth in zero gravity was harder than surviving at sea. Arranging facilities, finding water, and getting rid of the trash all became complex parts of the mission.

Space walking

One of the most dangerous things astronauts can do is leave the safety of a spacecraft. Out in space, they are exposed to all kinds of hazards: lack of air, radiation, extreme temperatures, and fast-moving space debris. However, space walking is essential—it enables astronauts to repair equipment, install new hardware, and even walk on the Moon.

▲ AIRLOCK *Astronauts enter space through a special room called an airlock. This room is sealed off from the rest of the spacecraft.*

FANCY A WALK?

During the early days of space exploration, Russia and the US were fierce rivals. When NASA announced that Ed White would soon make the first space walk, Russia decided to beat them to it, sending cosmonaut Alexei Leonov out on a space walk in 1965. The mission almost ended in disaster when Leonov's suit ballooned outward and he couldn't fit back through the door of the spacecraft. Only by reducing the pressure in the suit—a very dangerous thing to do—was he able to squeeze back into the airlock.

▲ ED WHITE *was the first astronaut to use jet propulsion during a space walk.*

▲ SPACE WALK
Astronauts Carl J. Meade and Mark C. Lee testing a SAFER jet pack 150 miles (240 km) above Earth in 1994.

▲ ROBOTICS
Mark C. Lee is shown anchored to the Remote Manipulator System (RMS) robotic arm on space shuttle Discovery.

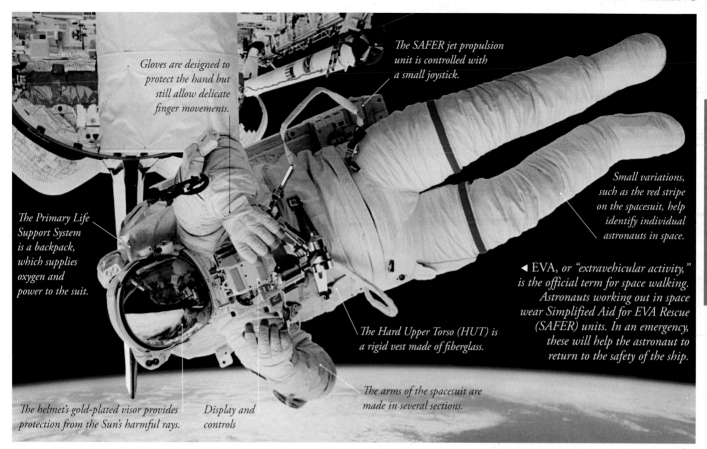

Gloves are designed to protect the hand but still allow delicate finger movements.

The SAFER jet propulsion unit is controlled with a small joystick.

The Primary Life Support System is a backpack, which supplies oxygen and power to the suit.

Small variations, such as the red stripe on the spacesuit, help identify individual astronauts in space.

◄ EVA, or "extravehicular activity," is the official term for space walking. Astronauts working out in space wear Simplified Aid for EVA Rescue (SAFER) units. In an emergency, these will help the astronaut to return to the safety of the ship.

The Hard Upper Torso (HUT) is a rigid vest made of fiberglass.

The arms of the spacesuit are made in several sections.

The helmet's gold-plated visor provides protection from the Sun's harmful rays.

Display and controls

Flying free

One of the greatest threats to space walkers is the possibility that they will accidentally drift away from the spacecraft, unable to return. The result would be a long, slow death in the emptiness of space. Nearly all space walkers are carefully tethered to the spacecraft, although special "flying armchairs" or jet packs are sometimes used, which allow astronauts to fly freely.

► SPACE ARMCHAIR
The Manned Maneuvering Unit (MMU) was used on three NASA missions in 1984.

Repair and construction

Crews working in space rely on handholds fitted on the outside of the spacecraft to move around. They may also be lifted to worksites by a robotic crane operated by another astronaut from inside the shuttle or space station. Lights on the spacesuit helmets allow astronauts to work in the dark.

► RECORD-BREAKER
On October 18, 2019, US astronaut Christina Hammock Koch completed the first ever all-female space walk (with fellow astronaut Jessica Meir), replacing a failed power control unit outside the International Space Station (ISS). Koch's 328-day stay on the ISS made history as the longest continuous spaceflight by a female astronaut.

Satellite recovery

In 1984, Manned Maneuvering Units (MMUs) were used to retrieve two faulty satellites that had become stuck in the wrong orbits. Astronauts Joe Allen and Dale Gardner performed an EVA, using the MMUs to reach the satellites and drag them back to the shuttle. The satellites were then returned to Earth for repairs. This was the last mission to use the MMU, which was retired by NASA soon after due to fears over its safety.

▲ ON APPROACH Dale Gardner moving toward satellite Westar VI.

▲ MANEUVERS Gardner and Allen guiding Westar to the shuttle.

Living in space

Sending people into space means providing the right conditions for them to live in. Between three and six people live on the International Space Station (ISS) at a time, usually staying for up to 6 months. The ISS is fitted with everything that the crew needs to make their mission comfortable and successful.

LEISURE TIME

When they are not busy working, astronauts on the ISS have many ways of relaxing. This includes spending time communicating with Earth by video-link, radio, or email. As well as chatting to friends and family, the crew speak with amateur radio enthusiasts and schools as they fly overhead.

▲ WINDOW ON THE WORLD *The ISS "cupola" is a large observation dome where astronauts can relax while looking down on Earth, as well as making scientific observations.*

▶ PLAYTIME *Many astronauts like to read, listen to music, watch DVDs, or play board games. Some play musical instruments— a keyboard, a guitar, and even a trumpet have been played in orbit.*

Keeping clean

The ISS crew cannot wash their hands under a faucet, like on Earth. Water does not flow in zero gravity, so there are no sinks or showers inside the station. When the astronauts want to get clean, they wipe themselves with alcohol or a wet towel containing liquid soap. Astronauts take sponge baths daily using two cloths—one for washing and one for rinsing. They use rinseless shampoo and usually swallow their toothpaste after brushing their teeth.

▲ SITTING COMFORTABLY *Astronauts strap themselves onto toilets that use suction to remove waste. On early missions, astronauts collected their waste in hoses and plastic bags.*

TAKE A LOOK: DOWN AT THE GYM

The human body loses muscle and bone in weightlessness, so to keep their muscles in shape, astronauts onboard the ISS go to the gym twice a day for an hour-long session of exercise. This ensures the astronauts do not collapse when they return to normal gravity. There are different exercise machines on the ISS, including a floating treadmill, exercise bikes, and an apparatus for "lifting" weights. The astronauts have to strap themselves onto the machines so they don't float away. The latest equipment enables the crew to perform resistance exercises (such as bench presses, sit-ups, and squats) despite the station's zero-gravity environment.

▶ UNEXPECTED SWELLING
In space, the human body's circulatory (blood) system turns upside down. Without gravity tugging the body's fluids downward, blood pressure is equal all over the body, so blood builds up in the head and causes swelling. Exercise helps to relieve this swelling.

Blood forced down

Blood spreads around body

Gravity on Earth

No gravity in space

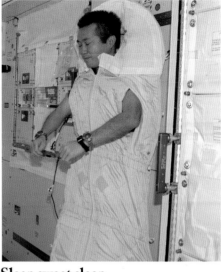

Sleep sweet sleep
Astronauts are happy sleeping almost anywhere—floor, wall, or ceiling—but they need to be near a ventilator fan. Without airflow, the carbon dioxide they breathe out will build up around them, leaving them gasping for oxygen.

Is it bedtime yet?
With 16 sunrises and sunsets a day on the ISS and the space shuttle, it's not easy to work out when it is time to sleep. Work schedules and sleep periods are based on the time at the mission control center in Houston, Texas.

FOOD AND DRINK

■ The first astronauts had to eat bite-sized cubes, freeze-dried powders, and pastes that were squeezed straight from a tube into the mouth!

■ Today, the ISS menu includes more than 100 different meals, plus snacks and hot and cold drinks. A lot of the food is freeze-dried, and water must be added before it can be eaten. All food is processed so that it does not have to be stored in a fridge.

▲ FOOD TUBE *The first space meals were soft, gloopy foods a lot like baby food.*

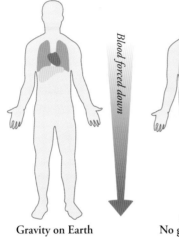
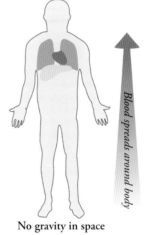
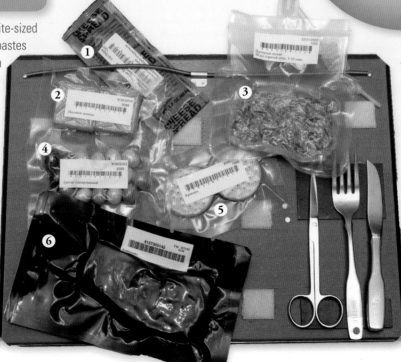

◀ A SOLID MEAL *Solid food can be eaten with a knife and fork, which are held down by magnets to stop them from floating away from the table.*

1 *Cheese spread*
2 *Shortbread cookies*
3 *Creamed spinach*
4 *Sugar-coated peanuts*
5 *Crackers*
6 *Beef steak*

Animals *in space*

Long before the first person set foot in space, scientists sent animals into orbit to see how they would cope with effects such as zero gravity. If animals could survive the journey into space, then maybe people could, too.

SPACE DOGS TAKE THE LEAD

Can canine success pave the way for human spaceflight?

▲ DOG DAYS *In 1960, Strelka and Belka (left) became the first animals in orbit to return to Earth alive. In 1966, Veterok and Ugolyok (above) spent 22 days in space. Their record stood until 1973.*

Laika the cosmonaut

Laika was the first animal ever to be sent into orbit. Scientists believed that dogs would be good candidates for spaceflight because they can sit for long periods of time. Unfortunately, Laika did not survive, dying about 5 hours into the trip.

▲ LAIKA *A stray off the streets of Moscow, Laika was quickly trained and sent into orbit in Sputnik 2 in November 1957. It was a major achievement for the Soviet Union in the space race against the US.*

Champion chimps

Chimpanzees are our nearest animal relatives, so it made sense to send some into space as a trial run ahead of humans. Many were trained and, in 1961, Ham was chosen as the first chimp to go into space. Although the capsule lost some air pressure during the flight, Ham's spacesuit protected him. The only thing he suffered from the 16-minute flight was a bruised nose.

TIMELINE OF SPACE ANIMALS

1940s

1947
Fruit flies were sent on a suborbital flight on a US V-2 rocket.

1948–1950
Five US suborbital flights carried three monkeys and two mice to altitudes of 80 miles (130 km). The mice survived.

1950s

1951
On September 20, Yorick the monkey and 11 mice were sent to an altitude of 44 miles (72 km) on a US Aerobee rocket. Yorick was the first monkey to survive a flight to the edge of space.

1957
Laika the dog became the first animal to be sent into orbit.

1959
Able, a rhesus monkey, and Miss Baker, a squirrel monkey, became the first living beings to successfully return to Earth after traveling in space on a suborbital flight.

Monkey business

There are obvious problems when sending animals into space. How do they feed themselves? How can their behavior be controlled? Monkeys on the Cosmos missions were strapped into seats for their own protection. They had been trained to bite on tubes to release food and drink, and also to press levers when a light shone, which kept them mentally alert.

◄ ANIMAL CARRIER *In 1983, the Cosmos 1514 mission took two monkeys and 10 pregnant rats into orbit. The trip lasted 5 days.*

 TAKE A LOOK: EGGS

There have been a number of experiments on eggs in space. Quail eggs fertilized on Earth and incubated on the Mir Space Station in 1990 did hatch, although there were not as many as would have hatched on Earth.

▶ SPACE CHICKS *Unfortunately, the quails that hatched on Mir did not survive very long.*

Mission TARDIS

These creatures are tardigrades, tough invertebrates that seem almost indestructible on Earth. But how would they fare in space? Mission TARDIS, a European Space Agency experiment, showed them to be the first animals to survive the weightlessness and coldness of space. They not only survived being frozen, but could also cope with UV light 1,000 times stronger than on Earth.

Weightless webs

On Earth, a spider uses wind and gravity to construct its web. So how would a spider spin a web in space, where there is neither of these? Two spiders, called Anita and Arabella, were sent into space onboard the 1973 Skylab 3 mission to find out. Once they got used to being weightless, they were soon spinning near-perfect webs.

Spiders use their weight to work out the thickness of the web silk.

The experiment was designed by an American schoolgirl named Judith Miles.

Scientists used the information from this experiment to find out more about how a spider's central nervous system works.

1960s		1970s	1990s	2000s	2010s
1960 Dogs Strelka and Belka's day trip into space ended with a safe return to Earth by parachute.	**1961** Ham became the first chimp in space.	**1973** Arabella and Anita, common cross spiders, were taken up by Skylab 3.	**1990** Journalist Toyohiro Akiyama took some Japanese tree frogs to the Mir Space Station.	**2008** ESA's Mission TARDIS sent tardigrades 167 miles (270 km) into space.	**2019** China's Chang'e 4 mission landed a capsule of seeds and fruit fly eggs on the far side of the Moon.

EXTENDING THE HOUSE

Imagine having to make building repairs to
your home—while hovering in low Earth orbit
210 miles (340 km) above New Zealand! Tethered
by the thinnest of wires, two astronauts go out on
a space walk to attach a new truss segment to the
International Space Station.

The first space *stations*

If astronauts want to live and work in orbit for months or even years, then the cramped cabin of a spacecraft is not practical. They need a much larger structure, known as a space station.

◄ SALYUT 1 *was powered by solar panels and completed 2,800 orbits of Earth.*

SALYUT 1

The world's first space station was the Soviet Union's Salyut 1, launched in 1971. The largest of its three sections was the service module, which housed the fuel, oxygen, and water tanks with the main engine at the rear. The central section was the work and living area. At the front was the docking section. A three-man crew lived in the station for 22 days, but after that, Salyut 1 remained unoccupied and was lowered from orbit later that year.

FAST FACTS

■ The name Salyut (salute) was a tribute to Yuri Gagarin, the first man in space, who had died in 1968.
■ Two Salyut stations (3 and 5) were used to spy on Western rivals. An onboard camera took detailed pictures of Earth's surface, and the film was returned to Earth in a special capsule.
■ Salyut 3 carried a machine gun in case of attack by other spacecraft. It was modified to work in the vacuum of space.

SCI-FI STATIONS

The first story about a space station, called "The Brick Moon," was published in a magazine in 1869. By the early 20th century, wheel-shaped space stations were in fashion in science-fiction. In reality, all the space stations built so far have been made of modules that are launched separately, then joined together when they are in orbit. The size and weight limitations of rockets have meant that stations have to be built like giant building blocks, one piece at a time.

▲ IN THE STORY, *the brick moon was accidentally launched with people onboard.*

▲ THE WHEEL-SHAPED STATION *was made famous when it appeared in the 1968 film* 2001: A Space Odyssey. *Space scientists did seriously consider wheel-shaped stations in the 1950s.*

Skylab crashed to Earth in 1979

Skylab

Skylab was the US's first space station and one of the heaviest spacecraft ever placed in Earth orbit. It was in use from 1973 to 1974. Skylab lost one of its two main solar panels when it was damaged during launch. But three crews were able to visit, with missions lasting 28, 59, and 84 days. They performed astronomy experiments, X-ray studies of the Sun, remote sensing of Earth, and medical studies.

MIR

This was the successor to the Russian Salyut series of space stations. The first module was launched in 1986 and was soon occupied by two crew members. Six more modules were added over the next 10 years, including a docking module for use by the space shuttle.

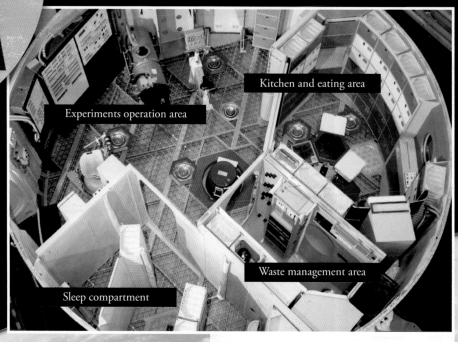

Kitchen and eating area

Experiments operation area

Waste management area

Sleep compartment

▲ SKYLAB WORKSHOP *This was the largest section of the space station. It contained the crew quarters, including a toilet, shower, and galley (kitchen), along with the laboratory facilities and a large waste disposal tank.*

▶ MIR IN ORBIT *The crew's quarters were in the base module. A service section contained the main engine and thrusters, while a third section housed five docking ports. In all, 31 crewed spacecraft and 64 cargo ships docked with Mir.*

Mir

Near disaster
In 1997, Mir suffered a serious fire. Four months later, an incoming Progress ship collided with the station, damaging the Spektr module and allowing air to leak into space. Luckily, the crew managed to close Spektr's hatch before they were forced to abandon the station and head for home.

The International *Space Station*

The International Space Station (ISS) is the largest and most expensive spacecraft ever built. Sixteen countries have worked together to construct and operate the station, and it will remain a permanent home in space for up to six astronauts at a time until at least the mid-2020s.

ISS facts and figures

- **Width (truss):** 356 ft (109 m)
- **Length (modules):** 290 ft (88 m)
- **Weight:** 925,000 lb (419,600 kg)
- **Operating altitude:** 240 miles (385 km) above Earth's surface.
- **Orbiting speed:** 5 miles (8 km) per second.
- **Atmospheric pressure inside:** 14.7 psi (1,013 millibars)—the same as on Earth.
- **Pressurized area:** 33,000 cubic ft (935 m³). This is about the same as a five-bedroom house.
- **Crew size:** Three to six people.

SOLAR POWER

The largest feature of the ISS is its eight pairs of solar panels. Each panel measures 240 ft (73 m)—longer than the wingspan of a Boeing 777 aircraft. The panels produce electricity from sunlight and can be turned so that they receive as much light as possible. They contain more than 262,000 solar cells, producing a maximum 110 kW of power.

First launches

At the core of the ISS are the Russian-built Zvezda and Zarya modules. Zarya was the first module to be launched into orbit, in 1998. It is now used mainly for storage and propulsion. The main living quarters were added in July 2000. America's Destiny, the first science lab, arrived in February 2001.

Working in the laboratory

Every day, ISS crews carry out science experiments in the labs. Hundreds of scientists on the ground also take part. These experiments cover many fields, including human biology, medical research, physical sciences, and Earth observation. Research topics range from growing protein crystals to making new metal alloys.

Robotic arms

The ISS has two robotic arms that are used to lift astronauts and pieces of equipment outside in space. The arms are controlled by astronauts inside the station. The main arm is called Canadarm 2, because it was built in Canada. It is 55 ft (16.7 m) long and can handle objects weighing up to 128 tons (116 tonnes)—the weight of a space shuttle. The arm has seven joints and four handlike grapple fixtures.

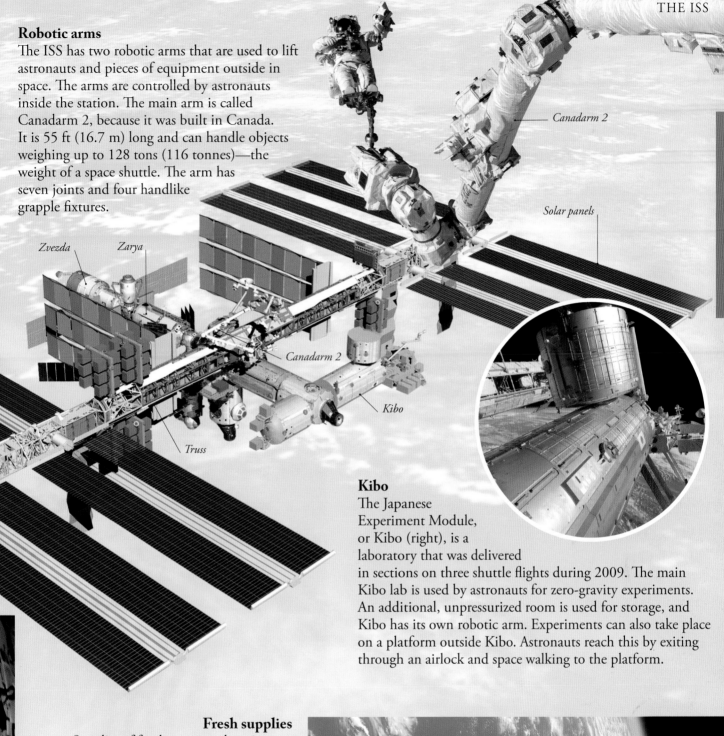

Canadarm 2

Solar panels

Zvezda

Zarya

Canadarm 2

Kibo

Truss

Kibo

The Japanese Experiment Module, or Kibo (right), is a laboratory that was delivered in sections on three shuttle flights during 2009. The main Kibo lab is used by astronauts for zero-gravity experiments. An additional, unpressurized room is used for storage, and Kibo has its own robotic arm. Experiments can also take place on a platform outside Kibo. Astronauts reach this by exiting through an airlock and space walking to the platform.

Fresh supplies

Supplies of food, water, and equipment are brought to the ISS by various uncrewed spacecraft. These include the Russian Progress cargo ship and Japan's H-II Transfer Vehicle, as well as two commercial craft—the SpaceX Dragon and Northrop Grumman's Cygnus. Since the space shuttle's retirement in 2011, astronauts have come and gone via Russian Soyuz vehicles, but from 2020, Boeing's new Starliner spacecraft and SpaceX's rival Crew Dragon plan to carry crew to and from the ISS.

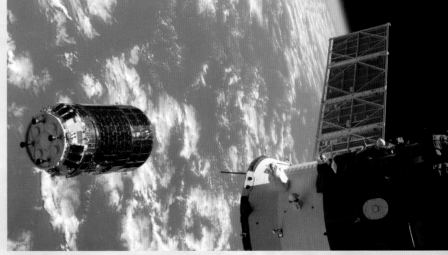

Science in *space*

The "microgravity" conditions of space offer a special environment for scientific research. Short periods of weightlessness can be created inside very tall drop towers or on aircraft flying high above Earth. However, the only way to experience weeks or months of weightlessness is onboard a space station.

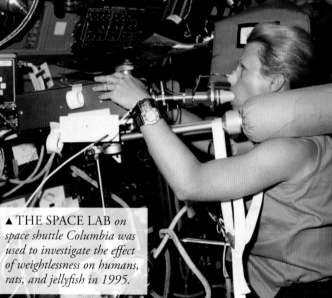

▲ GLOVEBOX EXPERIMENTS *Astronauts study the effects of zero gravity in the Destiny laboratory on the ISS. A glovebox provides a safe, enclosed area for experiments.*

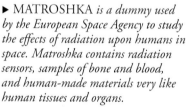

▶ MATROSHKA *is a dummy used by the European Space Agency to study the effects of radiation upon humans in space. Matroshka contains radiation sensors, samples of bone and blood, and human-made materials very like human tissues and organs.*

IMPROVING HEALTH

Without any gravity to push against, human muscles and bones become very weak. Astronauts on the International Space Station test ways of preventing damage to muscles and bones. This includes use of exercise machines, drugs, and small electric shocks.

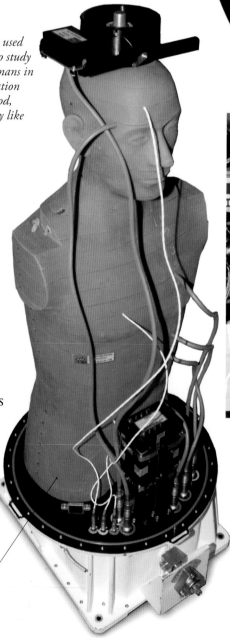

Matroshka is named after the famous Matryoshka Russian dolls because it is made of so many layers.

▲ THE SPACE LAB *on space shuttle Columbia was used to investigate the effect of weightlessness on humans, rats, and jellyfish in 1995.*

Space sickness

Many astronauts suffer from space sickness during their first few days in orbit. Since there is no up or down in space, the brain receives conflicting information from the eyes, muscles, skin, and balance organs. Numerous experiments have been done to study how the human brain deals with these signals and how it adapts to weightless conditions.

LIFE IN SPACE

Experiments with many different forms of life have been conducted in space, ranging from spiders and fruit flies to tomatoes, fish, and quail. Harmful bacteria seem to thrive in zero gravity, while the human ability to fight infections becomes weaker. It is impossible to sterilize spacecraft completely, so the spread of bacteria could be very dangerous for astronauts on long missions.

Crystals

Crystals grown in space are much bigger and have fewer flaws than those on Earth. Scientists are especially interested in studying protein crystals in space. There are more than 300,000 proteins in the human body yet very little is known about most of them. Producing protein crystals of high quality can help us to work out their shape and structure—and learn about how they work in the body.

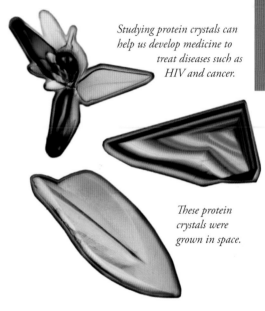

Studying protein crystals can help us develop medicine to treat diseases such as HIV and cancer.

These protein crystals were grown in space.

PLANTS IN SPACE

Plants were first sent into space onboard the Sputnik 4 satellite in 1960. Since then, scientists have been investigating how plants grow in space and looking at ways of growing lots of high-quality plants in a small area. This is important research for future space missions, on which astronauts might have to grow their own food, and also for growing crops on Earth.

▶ *THE ISS's VEGGIE experiment grows salad under special LED lights.*

Flames, liquids, and metals in space

Convection is the process by which hot liquids and gases on Earth rise and cool liquids and gases sink. Because convection can't operate in zero gravity, flames in space burn with a rounded shape rather than in the upward-pointing tapers seen on Earth. Liquids that would separate into layers of different densities on Earth also behave differently in zero gravity and mix very easily. Metals in liquid form can be mixed in space to form super-strong alloys that are much stronger than those made on Earth.

▲ EARTH FLAME
Flames on Earth point upward because heated air, which is less dense than the surrounding cooler air, rises up.

◀ SPACE FLAME
In zero gravity, convection has no effect, so flames burn with a rounded flame.

TAKE A LOOK: SPIN-OFFS

The transfer of technology from space use to everyday use is called a "spin-off." A lot of the science from space has found a use here on Earth.

■ **Golf-ball aerodynamics**
NASA technology was used to design a golf ball that would fly faster and farther when struck.

■ **Shock-absorbing helmet**
Protective helmets use a shock-absorbing padding first developed by NASA for use in aircraft seats.

■ **Fogless ski goggles**
A NASA-developed coating is applied to goggles, deep-sea diving masks, and fire helmets to prevent them fogging up.

■ **Quartz crystal**
NASA developed highly accurate clocks and watches using quartz crystal.

Space *tourism*

Nowadays, not everyone who goes into space is a professional astronaut. Scientists, various politicians, a Japanese journalist, two US teachers, and several businessmen have all flown. As space tourism becomes a reality, companies are springing up, offering to fly people on suborbital hops from new spaceports.

SPACESHIPONE

The race to build new types of spacecraft for tourists was boosted by a $10 million prize in 2004 from the X Prize Foundation. It was offered to the first company to build a spaceship that flew above 62 miles (100 km) twice within 2 weeks.

White Knight launcher

SpaceShipOne

▲ THE PRIZE *was won by SpaceShipOne, a three-seat research rocket built like an aircraft.*

Tourist rocket

The New Shepard, built by Jeff Bezos' Blue Origin company, is a reusable rocket designed to take a capsule with up to six passengers into space on a suborbital flight path.

◄ THE NEW SHEPARD *lifts off on an uncrewed test flight in 2016.*

FIRST TOURIST

■ The first person to pay for a flight into space was 60-year-old American Dennis Tito. The millionaire businessman went through a training program at Russia's Star City.

■ He flew on a Soyuz spacecraft to the ISS, arriving on April 30, 2001. He spent 6 days on the station before returning to Earth in another Soyuz.

■ While in space, Dennis Tito listened to opera, shot video and photos through the porthole, helped to prepare the meals, and spent time admiring the view as the space station swept around the planet once every 90 minutes.

SpaceShipTwo

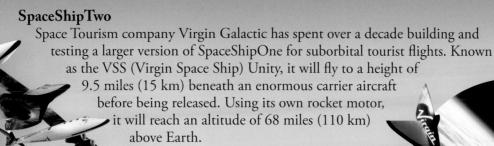

Space Tourism company Virgin Galactic has spent over a decade building and testing a larger version of SpaceShipOne for suborbital tourist flights. Known as the VSS (Virgin Space Ship) Unity, it will fly to a height of 9.5 miles (15 km) beneath an enormous carrier aircraft before being released. Using its own rocket motor, it will reach an altitude of 68 miles (110 km) above Earth.

SpaceShipTwo release

▶ TICKETS *for a trip on VSS Unity are priced at $250,000. The spacecraft will travel at Mach 3, faster than any fighter jet.*

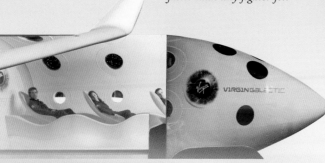

◀ THE CABIN *is 60 ft (18 m) long and 7½ ft (2.3 m) in diameter. It will carry six passengers and two pilots. Each passenger will sit by a large window and will be able to float freely for about 4 minutes before returning to Earth.*

Space hotels

Once cheaper ways of flying to space have been developed, space hotels are likely to be the next step. Ordinary people will then be able to orbit Earth and experience the wonders of weightlessness. Detailed plans have already been put forward for large inflatable modules in which people can stay. Once the first of these is launched, it can be joined by a propulsion unit and a docking module, enabling more inflatable sections to be added.

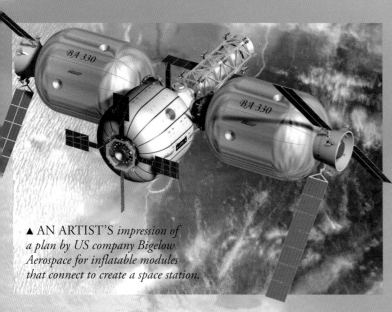

▲ AN ARTIST'S *impression of a plan by US company Bigelow Aerospace for inflatable modules that connect to create a space station.*

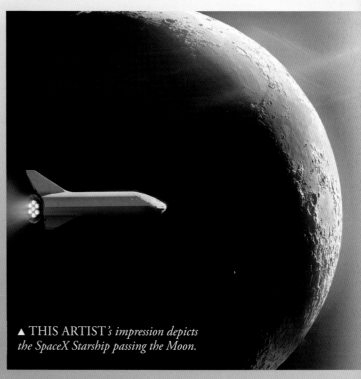

▲ THIS ARTIST*'s impression depicts the SpaceX Starship passing the Moon.*

Tourists to the Moon?

The commercial company SpaceX has ambitious plans for space tourism, taking passengers to the ISS, and future orbital hotels with its Super Heavy rocket. In 2017, it announced plans to send up to nine people on a trip around the Moon in its new Starship spacecraft, with a journey to Mars orbit planned for the mid-2020s.

Future *flyers*

Launchers have changed very little since the beginning of the space age more than 50 years ago—they still involve rockets and large amounts of heavy fuel. Space agencies are now trying to develop cheaper, reusable vehicles, but these would require new technologies, such as air-breathing engines.

One day, spacecraft may be able to reach orbit on a space elevator. Various designs have been proposed, usually involving a cable structure. This would stretch from the surface to geostationary orbit, with a counterweight at the upper end. Earth's rotation would keep the cable taut, so that a car or cabin could climb up the cable. Making this type of cable would require new materials that are light but strong.

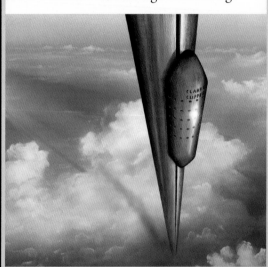

▼ SKYLON *This unpiloted, reusable spaceplane is being developed in the UK. It will carry engines that burn their fuel with air in Earth's atmosphere but switch to using an onboard "oxidant" chemical (like a normal rocket) in space.*

SPACEPLANES

This type of reusable vehicle is already being developed. A spaceplane has its own rocket engines and could one day carry people or cargo into orbit. It would take off from a runway or be carried to high altitude by an aircraft before being released. At the end of the mission, it would land on a runway, like an aircraft.

Commercializing space

Until recently, nearly all spacecraft delivering cargo and crew into space were developed by government space agencies, but this is now changing rapidly, with various commercial rocket companies competing for profitable satellite and crew launch contracts. New approaches and the urge to cut costs have led to major advances, especially when it comes to retrieving and reusing rocket stages. Companies such as SpaceX and Boeing are already providing cargo deliveries to the International Space Station, and crewed versions of their spacecraft will soon be taking American astronauts to orbit.

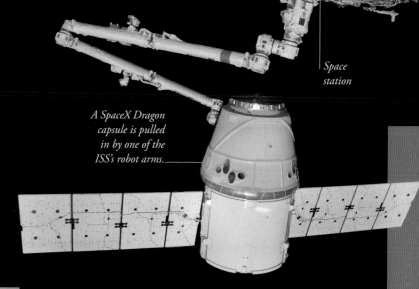

Space station

A SpaceX Dragon capsule is pulled in by one of the ISS's robot arms.

NASA's new rocket

With the space shuttle's retirement and the appearance of commercial spaceflight to orbit, NASA has been concentrating on a new rocket for exploration of the Moon and beyond. The Space Launch System (SLS) combines elements of the shuttle design into a stacked multistage rocket with the potential to send a conical Orion spacecraft all the way to the Moon or launch parts of a larger "deep-space" vehicle into Earth orbit for assembly.

Air-breathing launchers

Several countries are studying air-breathing engines, which would reduce the amount of liquid oxygen fuel that has to be carried. This type of launcher would be boosted to high speed by a normal jet engine or booster rocket. The engine, which has no moving parts, compresses air as it passes through, mixes it with fuel, and then ignites it.

TAKE A LOOK: POWER FROM SPACE

We are using more and more energy. With the threat of global warming, caused by the build-up of greenhouse gases, clean, renewable power is becoming increasingly important. One idea that is being studied is to get power from space using large solar panels flying in Earth orbit. The energy they generate could be beamed to the ground using lasers or microwaves and collected by gigantic dish antennas. The first Japanese test of space power could take place by 2030. Japan, China, and the US are all working on plans for space-based solar power plants.

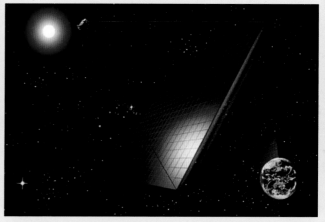

▲ GIANT SOLAR PANELS *orbiting above the equator could capture sunlight 24 hours a day and beam the energy to Earth.*

NASA's experimental X-43 set a still-unbroken aircraft speed record of 9.6 times the speed of sound in 2005.

113

Reaching *for the stars*

So far in the history of human space travel, 12 people have walked on the Moon and many more have lived aboard the International Space Station. One day, we may set foot on Mars and perhaps even settle on a planet in orbit around another star. But to do this, we have to overcome many challenges, including surviving the journey there.

A LONG JOURNEY

One of the main challenges of a crewed mission to Mars is the 6 months it will take to reach the planet, followed by a long stay, then the return trip. The crew of up to six people will be shut in a confined space, far from home. Messages will take up to 20 minutes to reach Earth, with the same delay for replies. They will have to learn to live together and deal with problems with little help from Earth.

ALONE WITH ALGAE

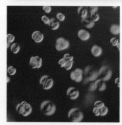

In the 1970s, the Russian BIOS 3 buildings in Siberia were used to test how people would cope with isolated living. Chlorella algae were grown indoors to recycle the air and make sure that the people living in the buildings didn't suffocate.

Chlorella algae

Living in isolation

A number of experiments have been carried out to see how people cope with isolation and cramped space. In the early 1990s, eight people were shut inside an artificial Earth during the Biosphere 2 project. The project lasted 2 years, and the biggest issues they faced were problems with the air system and arguments in the group.

📷 **WATCH THIS SPACE**

Biosphere 2 was built in Arizona. Different areas inside were built to mimic different environments on Earth. The view above is the ocean biome. Other biomes included grassland, rainforest, and desert.

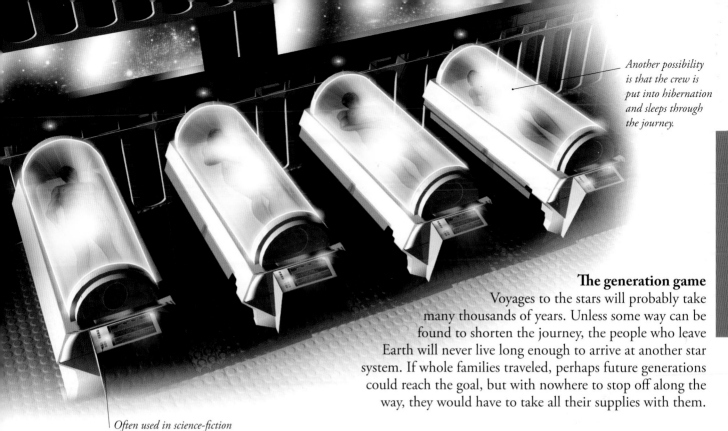

Another possibility is that the crew is put into hibernation and sleeps through the journey.

Often used in science-fiction films, human hibernation has yet to be achieved in real life.

The generation game

Voyages to the stars will probably take many thousands of years. Unless some way can be found to shorten the journey, the people who leave Earth will never live long enough to arrive at another star system. If whole families traveled, perhaps future generations could reach the goal, but with nowhere to stop off along the way, they would have to take all their supplies with them.

SPACE FARMING

'Yecora Rojo' 83 days old

Every 3 months, a cargo ship delivers a supply of food to the crew on the space station. These supplies are bulky, heavy, very expensive to deliver—and impossible to provide for a crew heading to Mars. A crew of six people would need 37,000 tons (33,000 tonnes) of food, water, and oxygen for a 3-year return trip. The answer is for astronauts to grow their own food. Experiments to grow plants from seed have already taken place in small space greenhouses.

FICTION AND REALITY

Science-fiction spaceships zip across the galaxy in next to no time, but real space travel is limited by the laws of physics. It is impossible to travel faster than the speed of light, so reaching the stars may take centuries of travel, no matter how much energy we have. Some NASA scientists, however, think it could one day be possible to bend space itself with a "warp drive," like that used by *Star Trek*'s USS Enterprise.

The USS Enterprise reaches warp speeds using engines powered by the strange (but real) substance known as antimatter.

RECYCLING

Scientists try to find ways to recycle as much waste as possible on spacecraft. There are already machines that purify urine for drinking and washing. Oxygen for breathing can be made by splitting water atoms. Systems are also being developed that use bacteria to recycle human waste for use in growing food and producing water.

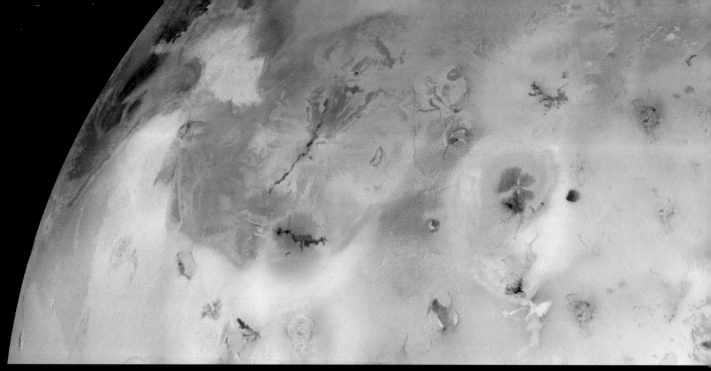

SOLAR
SYSTEM

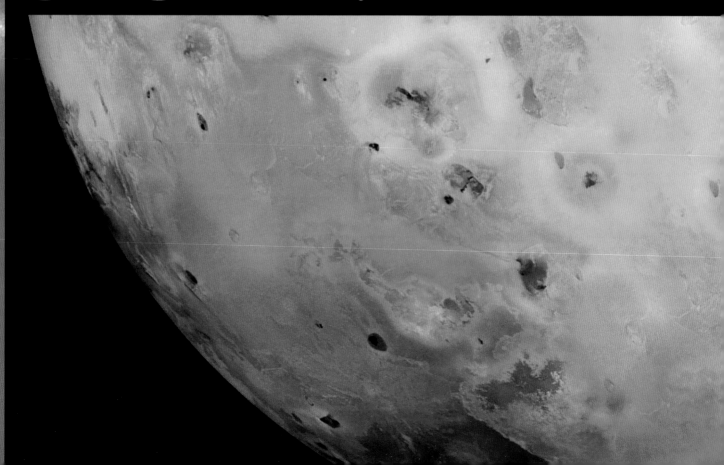

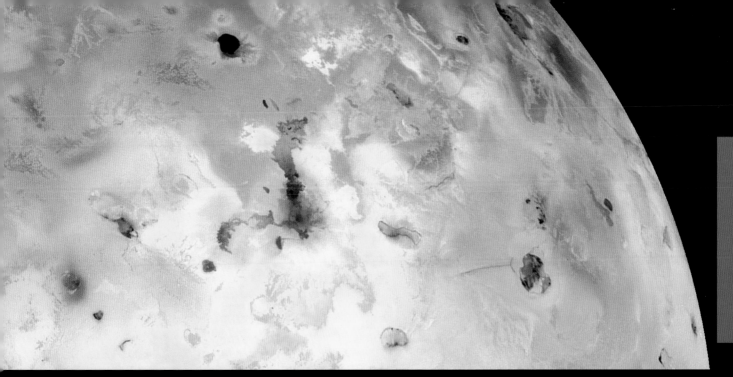

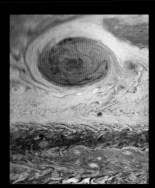

Our solar system is the region of space that falls under the influence of the Sun's gravity. It extends two light-years into space and contains planets, moons, asteroids, and comets.

Birth of the *solar system*

Everything in the solar system—the Sun, planets, moons, and smaller objects—was born inside a vast, spinning cloud. The story began about 5 billion years ago, with a giant cloud made mostly of hydrogen gas and dust. The cloud began to shrink and contract. Eventually our Sun formed in the center of the cloud, where it was denser and hotter. The rest of the cloud formed a swirling disk called the solar nebula.

◀ SHOCK REACTION
No one knows why the cloud began to shrink, but it may have been triggered by a shock wave from a star that exploded as a supernova.

COLLISIONS AND MERGERS

As the planetesimals grew bigger, their gravity pulled more material toward them, which led to more collisions. Eventually, regions of the nebula were dominated by a few large bodies. In the outer solar system, these objects attracted huge amounts of gas. This led to the formation of the planets known as gas giants—Jupiter, Saturn, Uranus, and Neptune.

THE SOLAR NEBULA

Within the solar nebula, dust and ice particles were colliding and merging. Through this process, the tiny particles grew into larger bodies a few miles across. In the inner, hotter part of the solar nebula, these building blocks (called planetesimals) were mostly made of rock and metals. Farther from the center, where the nebula was much colder, they were mainly made of water ice.

TAKE A LOOK: THE BIRTH OF THE MOON

Most scientists think that the Moon was born during a collision between a Mars-sized object and the young Earth. It may have taken only a few hundred years from the time of collision until the formation of the Moon. At first, the Moon was much closer to Earth than it is now, orbiting once every few days. Now it takes just over 27 days to complete one orbit.

▲ COLLISION COURSE
A large object, the same size as the planet Mars, collided with Earth.

▲ CRACKING UP *The impact vaporized and melted parts of Earth and the object, throwing debris into space.*

▲ ALL IN ORDER *The debris from the collision formed a ring around Earth.*

▲ NEW MOON *Material within the ring eventually combined to form our Moon.*

OTHER PLANETARY SYSTEMS

Planetary systems are now known to be very common. Most young stars in our galaxy are surrounded by disks of dust and hydrogen gas—just like the young Sun. By studying these stellar disks, scientists can learn a lot about the early history of the solar system. Thousands of planets have been found in orbit around distant stars. Most of the early discoveries were of large, Jupiter-type planets, but as techniques have become more sensitive, many planets the size of Earth and smaller are being found.(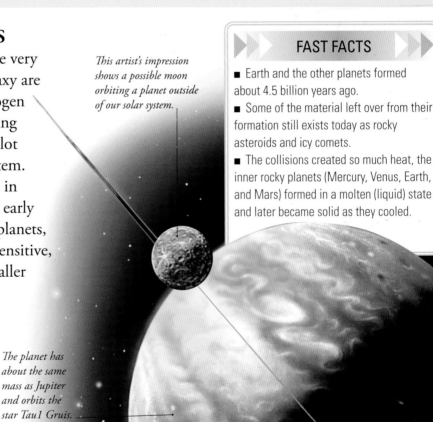 pp.226–227).

This artist's impression shows a possible moon orbiting a planet outside of our solar system.

The planet has about the same mass as Jupiter and orbits the star Tau1 Gruis.

FAST FACTS

■ Earth and the other planets formed about 4.5 billion years ago.
■ Some of the material left over from their formation still exists today as rocky asteroids and icy comets.
■ The collisions created so much heat, the inner rocky planets (Mercury, Venus, Earth, and Mars) formed in a molten (liquid) state and later became solid as they cooled.

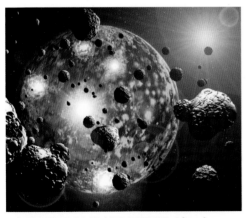

▲ HEAVY BOMBARDMENT *After the planets had formed, there was still a lot of material left over. Most of the fine material was blown away by a strong solar wind. Larger rocks continued to collide with Earth and the other planets until about 4 billion years ago.*

The Sun's *family*

The Sun rules over a vast area of space. Its gravity, radiation, and magnetic influence extend outward for billions of miles. Within this solar system are eight planets and their moons, at least five dwarf planets, millions of asteroids, and billions of comets.

MERCURY *The closest planet to the Sun has changed little in billions of years. It is a small, heavily cratered world with no atmosphere and no moons. Its year lasts 88 Earth days.*

PLUTO *Discovered by Clyde Tombaugh in 1930, Pluto was once known as the ninth planet from the Sun, but it is now classified as a dwarf planet.*

MARS *The fourth planet from the Sun has many craters, as well as volcanoes, rift valleys, and winding canyons. It also has two moons. Its year lasts 687 Earth days.*

URANUS *Discovered by William Herschel in 1781, the seventh planet from the Sun has a dark ring system and 27 moons. Its year lasts 84 Earth years.*

JUPITER *The fifth planet from the Sun is also the largest. It has thin rings, about 80 known moons, and a cloud feature called the Great Red Spot. Its year lasts almost 12 Earth years.*

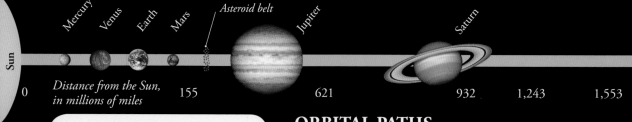

Sun | Mercury | Venus | Earth | Mars | Asteroid belt | Jupiter | Saturn | Uranus

0 | *Distance from the Sun, in millions of miles* 155 | 621 | 932 | 1,243 | 1,553

Inner planets
The four inner planets (Mercury, Venus, Earth, and Mars), asteroids, and many of the moons are made of rock. The rocky planets are much smaller than the gassy outer planets. They also have fewer moons (some have none at all) and no rings.

ORBITAL PATHS
Most of the planets, moons, and asteroids travel in almost circular orbits in the same direction (west to east) around the Sun. Most orbits also lie close to the plane of Earth's orbit, called the ecliptic. So if you looked at the solar system side-on, you would see most of the orbits are roughly on the same level. Mercury's and Pluto's orbits are not—they orbit at an angle.

Orbits and rotations

An orbital period is the time it takes one object to travel around another in a complete circuit. The orbital period of a planet around the Sun is also the length of its year. The rotational period of a planet is how long it takes to make a complete turn on its axis. This is its day.

ASTEROID BELT *Lying between Mars and Jupiter, the belt is around 112 million miles (180 million km) wide and contains a million or so asteroids.*

NEPTUNE *Discovered by Johann Galle in 1846, the eighth planet from the Sun has a thin ring system and 14 moons. Its year lasts almost 165 Earth years.*

VENUS *The second planet from the Sun is similar in size to Earth, but the air pressure is 90 times greater than on Earth. It has no moon. Its year lasts 224 Earth days.*

SATURN *The sixth planet from the Sun is the second largest, after Jupiter, but is light enough to float. It has 82 known moons, and its year lasts 29.5 Earth years.*

EARTH *The third planet from the Sun is the largest of the four rocky planets and the only planet with liquid water. Its year lasts 365 days.*

COMET HALLEY

The order of the planets

If you find it tricky to remember the order of the eight planets of the solar system, try using this sentence to help you: **M**y **V**ery **E**ducated **M**other **J**ust **S**erved **U**s **N**oodles (**M**ercury, **V**enus, **E**arth, **M**ars, **J**upiter, **S**aturn, **U**ranus, **N**eptune).

Neptune

| 1,864 | 2,175 | 2,485 | 2,800 |

Outer planets

The four large outer planets (Jupiter, Saturn, Uranus, and Neptune) are known as gas giants. This is because they are made of gases, with a solid core of rock and ice. The farthest objects from the Sun, such as Pluto and the comets, are mostly made of ice.

DWARF PLANETS

A dwarf planet is like other planets—it revolves around the Sun and reflects the Sun's light. However, a planet clears all other objects from its orbit, whereas there are still lots of objects in a dwarf planet's orbit. There are five known dwarf planets—Pluto, Eris (the largest), Ceres (also the largest asteroid), Haumea, and Makemake. They are icy debris left over from the formation of the planets 4.5 billion years ago.

◄ PLUTO *The best-known of the dwarf planets, Pluto is a dark, icy world with five moons and no atmosphere. It is smaller than Mercury, and its year lasts 248 Earth years.*

Mercury

Mercury is the smallest planet. It is also the closest planet to the Sun, so we always see it near the Sun in the sky. This makes it very hard to see, except at sunrise or sunset, because it is hidden by the Sun's glare. Mercury has no moons and is too small to hold onto any atmosphere.

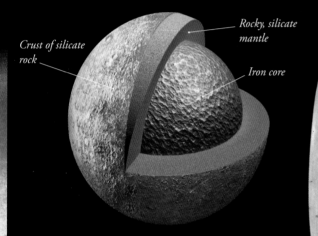

Crust of silicate rock

Rocky, silicate mantle

Iron core

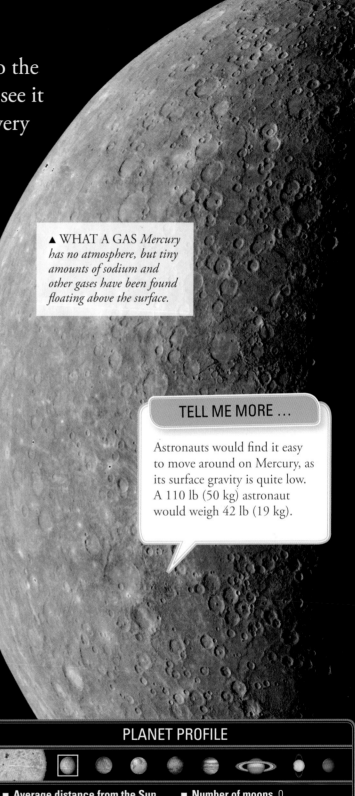

▲ WHAT A GAS *Mercury has no atmosphere, but tiny amounts of sodium and other gases have been found floating above the surface.*

TELL ME MORE …

Astronauts would find it easy to move around on Mercury, as its surface gravity is quite low. A 110 lb (50 kg) astronaut would weigh 42 lb (19 kg).

A SMALL WORLD

Mercury is very small—about 18 Mercurys would fit inside Earth. But it is denser than any planet except Earth. This is because it seems to have a very large core of iron and nickel, covered by a rocky mantle and crust. Mercury's iron core produces a magnetic field that is 100 times weaker than Earth's. This may be because Mercury spins more slowly on its axis.

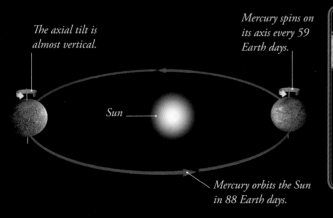

The axial tilt is almost vertical.

Mercury spins on its axis every 59 Earth days.

Sun

Mercury orbits the Sun in 88 Earth days.

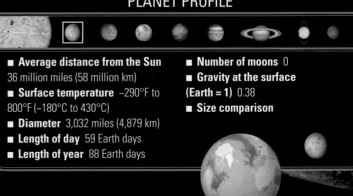

PLANET PROFILE

- **Average distance from the Sun** 36 million miles (58 million km)
- **Surface temperature** −290°F to 800°F (−180°C to 430°C)
- **Diameter** 3,032 miles (4,879 km)
- **Length of day** 59 Earth days
- **Length of year** 88 Earth days
- **Number of moons** 0
- **Gravity at the surface (Earth = 1)** 0.38
- **Size comparison**

A giant meteorite strikes Mercury, forming the Caloris Basin.

Shock waves travel through the core …

… and spread over the surface.

Shock waves meet and shatter the surface opposite the impact site.

Giant impact basins

Like the Moon, Mercury is covered with craters. These show that it has been battered by millions of impacts with asteroids and meteorites since it formed. Some of these impacts blasted out huge hollows in the surface. The most famous of these is the circular Caloris Basin, which is about 930 miles (1,500 km) across. Its floor shows ridges and fractures, with mountains around the edge. The explosion that formed the Caloris Basin seems to have sent shock waves through the planet. These produced a large area of jumbled hills on the opposite side of Mercury.

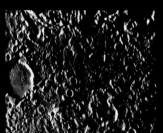

◄ CHAOTIC TERRAIN *Land opposite the Caloris Basin shows the impact of shock waves, which have caused fault lines, small lines, and depressions.*

TAKE A LOOK: TRANSIT ACROSS THE SUN

Mercury is the closest planet to the Sun, although its orbit is more oval-shaped (elliptical) than circular, so it varies from 28 million miles, or less than one-third Earth's distance (46 million km), to 44 million miles, or almost half Earth's distance (70 million km). Sometimes Mercury passes exactly between the Earth and the Sun. We see the planet as a tiny dot moving slowly across the face of the huge Sun. This is known as a transit, and it can only happen in May or November. The next transit of Mercury will be on November 13, 2032.

▶ MERCURY'S JOURNEY *On the evening of November 8, 2006, Mercury moved across the Sun. It finished its journey just after midnight. The three tiny black dots show how small Mercury is compared to the Sun.*

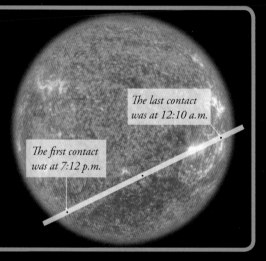

The last contact was at 12:10 a.m.

The first contact was at 7:12 p.m.

MESSENGER

Sunshade

Large velocity adjust (LVA) thruster

Star trackers

Helium tank

Solar panel

Magnetometer

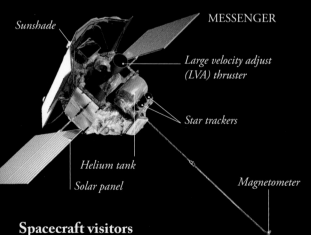

Spacecraft visitors

Mercury's rapid orbit makes it very hard for spacecraft to reach. The Mariner 10 mission made three brief flybys of the planet in 1974 and 1975, but it took decades before a probe made it to orbit. The MESSENGER mission orbited the planet for 4 years from 2011 before being deliberately crashed onto its surface. Another spacecraft, called BepiColombo, should arrive at Mercury in 2025.

The equatorial region nearest the Sun is the hottest area.

Hot and cold spots

The sunlit side of Mercury gets very hot, especially close to the equator where the Sun is overhead and sunlight is most intense. The Caloris Basin lies in one of these hot spots—Caloris is Latin for "heat." Temperatures here can reach 800°F (430°C)—hot enough to melt lead. Despite the intense heat, there is evidence that water ice may exist at the bottom of deep craters near the planet's poles.

An astronaut would fry in the heat of the day.

There is no air to spread heat, so Mercury's night side is very cold.

Venus

Venus is the most similar planet in the solar system to Earth. Although it is closer to the Sun, making it hotter than Earth, both planets are similar in size, mass, and composition. However, Venus has no water or life and is covered with a very thick, suffocating atmosphere.

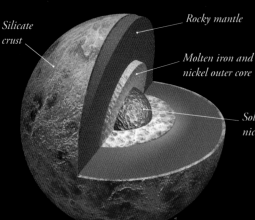

Silicate crust

Rocky mantle

Molten iron and nickel outer core

Solid iron and nickel inner core

▶ ROCKY TERRAIN
The highest mountains on Venus are the Maxwell Montes. They rise almost 7.5 miles (12 km) above the ground and are taller than Mount Everest.

DON'T GO THERE!

Venus may be closer to Earth than any other planet, but you wouldn't want to go there. Thick clouds of sulfuric acid and a suffocating blanket of carbon dioxide gas trap the Sun's heat, turning it into a scorching oven. Astronauts visiting Venus would die from a combination of acid burns, roasting, crushing, and suffocation.

About 80 percent of sunlight reflects away.

Thick clouds of sulfuric acid stop most of the sunlight reaching the surface.

Reflected light makes the cloud surface bright and easy to see.

Carbon dioxide in the atmosphere absorbs heat so it cannot escape.

Just 20 percent of sunlight reaches the surface.

PLANET PROFILE

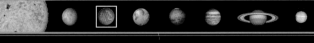

- **Average distance from the Sun** 67 million miles (108 million km)
- **Surface temperature** 865°F (460°C)
- **Diameter** 7,521 miles (12,104 km)
- **Length of day** 243 Earth days
- **Length of year** 224.7 Earth days
- **Number of moons** 0
- **Gravity at the surface (Earth = 1)** 0.91
- **Size comparison**

Cloud cover

The surface of Venus is hidden by a dense layer of pale yellow clouds. These are made of sulfur and sulfuric acid. Winds move the clouds around the planet from east to west at about 220 mph (350 kph). This wind sweeps the clouds right around Venus in only 4 days.

Spinning around

Venus spins very slowly clockwise, the opposite of most other planets. If you were standing on Venus, you would see the Sun go backward across the sky, rising in the west and setting in the east. It takes 243 Earth days to rotate once, so its day is longer than its year (224.7 Earth days).

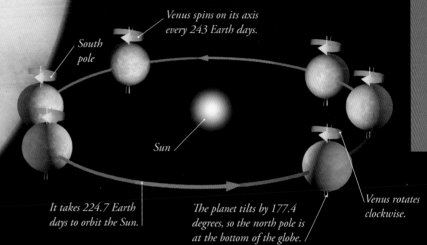

Venus spins on its axis every 243 Earth days.

South pole

Sun

It takes 224.7 Earth days to orbit the Sun.

The planet tilts by 177.4 degrees, so the north pole is at the bottom of the globe.

Venus rotates clockwise.

📷 WATCH THIS SPACE

There are more than 1,600 volcanoes on Venus. Among the more unusual features are the pancake lava domes, up to 40 miles (65 km) across and 3,300 ft (1,000 m) high. They are probably small eruptions of very thick, sticky lava that flowed onto a flat plain and then cooled before it could flow very far.

🔍 TAKE A LOOK: THE SWIRLING SOUTH

The first ever image of Venus's south pole was taken by the European Space Agency's Venus Express in 2006. Taken from more than 124,000 miles (200,000 km) away from the planet, this shows the "night side" of Venus (the hemisphere that is away from the Sun). It was taken by a VIRTIS spectrometer, which uses heat as well as light to make images. False color added to the picture shows clouds swirling around the south pole.

The darker the red, the thicker the clouds.

Brighter red shows thinner cloud, where heat has escaped and been picked up by VIRTIS.

There is a double vortex over the south pole. This is the center of the spinning clouds.

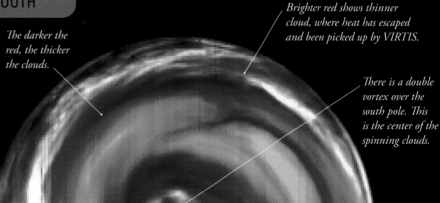

Touchdown

In March 1982, the Venera 13 and 14 landers sent back the only color pictures we have from the surface. They showed an orange sky and a desert covered in rocks of different sizes. Many of these were flat, suggesting thin layers of lava. At least 85 percent of the surface of Venus is covered in volcanic rock.

▶ VENERA ON VENUS *Venera 13 and 14 carried soil samplers to test the surface of Venus.*

CCCP

Views of Venus

As our closest neighbor, Venus is an obvious place to send space probes. The first successful landing was in 1970—all the earlier probes were destroyed by the extreme heat and pressure. Since 1978, orbiters have used radar to peer through the thick cloud and reveal the surface.

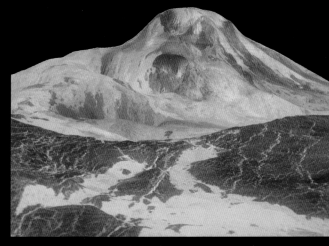

▲ COMPUTER DESIGN *This computer-created image of Maat Mons is based on radar data from the Magellan orbiter. Colors are based on images returned by the Venera 13 and 14 landers.*

VENUSIAN VOLCANOES

The most noticeable features on the surface of Venus are its volcanoes, of which there are at least 1,600. The tallest is Maat Mons (the peak at the back of the landscape below), about 3 miles (5 km) high. Its lava flows stretch for hundreds of miles across the surrounding plains.

It's thought that Maat Mons is not currently active, but no one knows for sure.

▲ DOUBLE SUMMIT *The Magellan spacecraft used radar to get this image looking straight down on Sapas Mons. The two dark spots are its mesas (flat tops).*

Sapas Mons

This landscape is the Atla Regio, a region in the northern hemisphere of Venus that was probably formed by large amounts of molten rock rising up from inside the planet. The bright area to the front is Sapas Mons, a shield-shaped volcano 135 miles (217 km) across that gently rises to a height of 1 mile (1.6 km) above the surrounding terrain.

▶ LINE UP *The Ovda Regio area of Aphrodite Terra is crossed by long, narrow ridges. The dark patches may be lava or wind-blown dust.*

Aphrodite Terra

Just as there are mountains and plains on Earth, Venus has highlands and lowlands, too. The largest highland region is Aphrodite Terra, in the equatorial area of Venus. The size of a continent on Earth, it runs two-thirds of the way around Venus and is divided into two main regions: the western Ovda Regio and the eastern Thetis Regio.

▼ THREE CRATERS *Magellan found this trio of craters in the Lavinia Planitia region of Venus. The distance between them is no more than 340 miles (500 km).*

▲ CRATER CREATOR *Combining radar data from Magellan and color images from Venera 13 and 14, we can see how Howe Crater appears on Venus. It is 24 miles (38 km) wide.*

Impact craters

Compared to other planets, Venus doesn't appear to have many impact craters. This might be because many meteorites burn up in the thick atmosphere before they reach the surface and create an impact crater. Another idea is that the surface of Venus is too young to have had many collisions with large meteorites. Most of the craters on the planet are less than 500 million years old.

🎥 WATCH THIS SPACE

Maxwell Montes are the highest mountains on Venus, rising over 6 miles (10 km). The color suggests that the rock is rich in iron.

Pioneer Venus

NASA's Pioneer Venus mission was made up of two different spacecraft. The orbiter, launched in 1978, was the first spacecraft to use radar to map the surface. It burned up after 14 years. Pioneer Venus 2 carried four probes to collect atmospheric data.

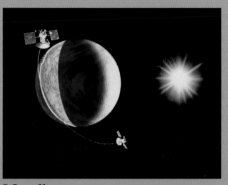

Magellan

Launched in May 1989, NASA's Magellan spacecraft arrived at Venus in August 1990. It spent more than 4 years in orbit and produced the most detailed radar map of the planet's surface. It was deliberately burned up in Venus's atmosphere in 1994.

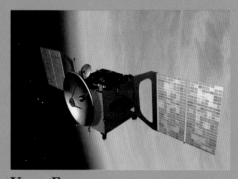

Venus Express

Europe's first mission to Venus was launched in November 2005, arriving at Venus in April 2006. It flew over the planet's polar regions, studying the cloud layers and atmosphere in great detail from the cloud tops to the surface. Venus Express burned up in 2014.

Mars

After Earth, Mars is the most suitable planet for humans to live on. Its day is only a little over 24 hours long, and it has Earth-like seasons. Mars was named after the Roman god of war because of its blood-red color, which is caused by rusty iron-rich rocks.

TELL ME MORE …

Visitors to Mars would have to wear spacesuits in order to breathe. The air is very thin and mainly carbon dioxide, a suffocating gas.

Red sky at night

The Martian sky is full of fine dust, which makes it appear orange-red. It means that sunsets on Mars are always orange-red, and there's so much dust, the sky stays bright for an hour after sunset. The daytime temperature can reach a pleasant 77°F (25°C) in summer, but it plummets as soon as the Sun sets and can drop to a bitter –195°F (–125°C) on winter nights.

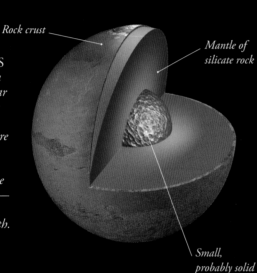

Rock crust

Mantle of silicate rock

Small, probably solid iron core

▶ MINI MARS
The surface area of Mars is similar to that of all the continents on Earth. Details are hard to see from ground-based telescopes because Mars is so small—about half the diameter of Earth.

POLAR ICE CAPS

There are permanent ice caps at both Martian poles, but they are quite different. The northern ice sheet is 1.2 miles (2 km) thick and mainly made of water ice. The southern polar cap is thicker and colder (–166°F/–110°C even in summer) and contains more carbon dioxide ice.

MARTIAN MOONS

Mars has two small, black, potato-shaped moons called Phobos and Deimos. They may be asteroids that were captured by Mars long ago. Phobos is slightly larger than Deimos and has a large impact crater called Stickney. Both are heavily cratered and seem to be covered in a layer of dust at least 3 ft (1 m) thick.

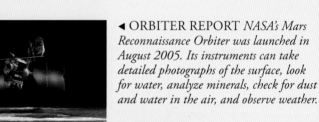

◀ ORBITER REPORT *NASA's Mars Reconnaissance Orbiter was launched in August 2005. Its instruments can take detailed photographs of the surface, look for water, analyze minerals, check for dust and water in the air, and observe weather.*

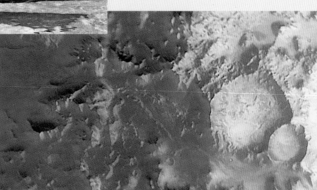

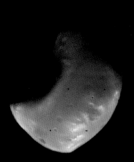

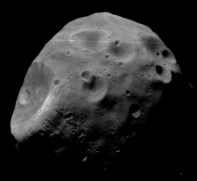

Deimos completes one orbit of Mars every 30 hours.

Phobos is much closer to Mars, completing one orbit every 7 hours 40 minutes.

▲ MOUNTAIN FROST *Mars changes in appearance with the seasons. Here, a sprinkling of carbon dioxide frost gives a wintry look to Charitum Montes, a mountain range in the planet's southern hemisphere.*

TAKE A LOOK: DUST STORMS

Mars is a dry planet, although there is lots of evidence that there used to be water on its surface. Today, the temperature is too cold and the air too thin for liquid water to exist on the surface. But the planet does have lots of wind. High-level winds reach speeds of up to 250 mph (400 kph), kicking up huge clouds of dust 3,000 ft (1,000 m) high. The dust storms can cover vast areas of the planet and may last for months.

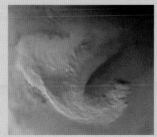

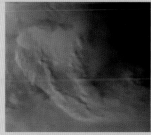

▲ BREWING A STORM
The beginning of a storm takes shape on June 30, 1999.

▲ SHAPE SHIFTER
A cloud of orange-brown dust is raised by high winds.

▲ GETTING LARGER
Dust blows over the northern polar ice-cap (the white area in the top middle of the image).

▲ ... AND LARGER STILL
This photo was taken 6 hours after the first one, and the storm is still building.

The heights of Olympus

Mars has the largest volcanoes in the solar system. The most impressive is called Olympus Mons, or Mount Olympus. At 373 miles (600 km) across, it would cover most of England, and at 13 miles (21 km) high, it is over twice as tall as Mount Everest. In the center is a huge, sunken crater that is 56 miles (90 km) across.

Olympus Mons is the largest volcano in the solar system.

Viking 1 Lander and Pathfinder landed near the Chryse Planitia.

In places, the Kasei Vallis valley is more than 2 miles (3 km) deep. It was the result of a devastating flood.

The volcanoes Ascraeus Mons, Pavonis Mons, and Arsia Mons make up the Tharsis Montes range.

Valles Marineris runs like a scar just below the Martian equator. This system of canyons is 2,485 miles (4,000 km) long.

The crater Lowell is 4 billion years old.

PLANET PROFILE

- **Average distance from the Sun**
142 million miles (228 million km)
- **Surface temperature** −195°F to 77°F (−125°C to 25°C)
- **Diameter** 4,220 miles (6,792 km)
- **Length of day** 24.5 hours (1 Earth day)
- **Length of year** 687 Earth days
- **Number of moons** 2
- **Gravity at the surface (Earth = 1)** 0.38
- **Size comparison**

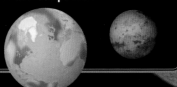

Mars missions

We know more about Mars than any other planet (except Earth). More than 25 spacecraft have been sent to study it since 1965, and the number of missions is increasing every couple of years as more robots are sent up. Eventually, these missions may pave the way for human colonization of the planet.

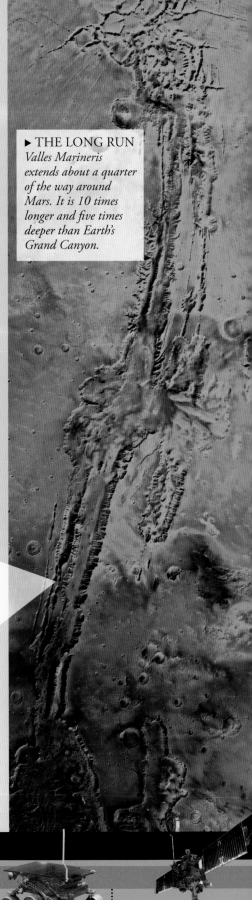

▶ THE LONG RUN
Valles Marineris extends about a quarter of the way around Mars. It is 10 times longer and five times deeper than Earth's Grand Canyon.

WATCH THIS SPACE

These gullies, or channels, run down from cliffs (top left) into a crater. They look like those on Earth that have been carved out by flowing water.

WHY EXPLORE MARS?

Mars is the most Earth-like planet in the solar system and one of the closest planets to Earth. As missions landed on the surface, we learned more about Mars, including finding lots of evidence that there was once liquid water on Mars. Now the search is on for signs of life.

Geography and geology

The valleys, volcanoes, and other surface features on Mars were formed in one of three ways: by tectonics (movement of the planet's crust); by water, ice, or wind; or by meteorite impacts. The largest tectonic feature is Valles Marineris, running like a huge gash across the planet. This series of canyons was created billions of years ago, when the surface of the young planet was stretched and split by internal movement.

▲ LOTS OF LAYERS *The above image shows the floor of one of the chasmata, or canyons, in the Valles Marineris. The floor is made up of about 100 layers of built-up rock.*

IMPORTANT MARS MISSIONS

1960s

1964
Mariner 4 (US) made the first successful flyby, taking 21 images.

1970s

1971
Mariner 9 (US) became the first successful Mars orbiter.

1976
Viking 1 (US) made the first successful landing on Mars.

1990s

1997
Mars Pathfinder (US) delivered the first successful rover to Mars.

1997
Mars Global Surveyor (US) mapped the entire planet, providing more evidence that water had flowed on Mars in the past.

► RED PLANET
This true-color view of Endurance Crater was taken by the Opportunity rover as it stood on the western rim.

Endurance Crater

When large meteorites crash-land, they leave impact craters (👁 pp.160–161). Endurance Crater is quite small—about 420 ft (130 m) wide and no more than 100 ft (30 m) deep. Around the crater are small, dark gray pebbles that scientists nicknamed "blueberries." They contain an iron-rich mineral called hematite. On Earth, hematite forms in lakes and springs, so the pebbles could be a sign of water on Mars.

▲ DUSTY DUNES *The middle of the crater's floor looks like a desert. Red dust has piled up into small sand dunes that are up to 3 ft (1 m) tall.*

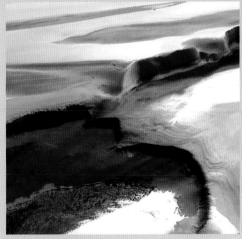

Polar ice cap

Like Earth, Mars has ice caps at its northern and southern poles. The caps can be seen from Earth, but missions to Mars allow scientists to study them closely. In winter, the ice is covered in frozen carbon dioxide. In summer, this evaporates and just the caps of water ice remain.

► A GOOD OPPORTUNITY
In 2004, the Opportunity rover spent 6 months taking images and examining rocks and soil in Endurance Crater. The rover finally lost contact with Earth in 2018.

▼ PROBING MARSQUAKES
In 2018, NASA's InSight lander arrived on Mars. The mission carries seismometers (earthquake sensors) that will help reveal the structure of the Martian interior.

Climbing Mount Sharp

NASA's Curiosity rover landed in Gale Crater in 2012. This crater is thought to be an ancient lake bed and has a large peak at its center called Mount Sharp, which Curiosity has been investigating since 2014.

2000s

2003
Europe's Mars Express orbiter began taking detailed pictures of Mars.

2008
Phoenix (US) landed in Martian Arctic and operated for over 5 months (before its batteries went flat).

2010s

2012
NASA's Curiosity rover landed and began exploration of Gale Crater.

2014
India's Mars Orbiter Mission arrived in orbit, carrying five science instruments.

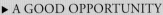

MARTIAN SAND ART

This close-up image from the Mars Reconnaissance Orbiter looks like an elaborate tattoo, but it's actually sand on the surface of the planet. The patterns have been created by dust devils—spinning columns of rising air up to 5 miles (8 km) high. As they whirl across the surface of Mars, they pick up loose red dust, uncovering darker, heavier sand underneath.

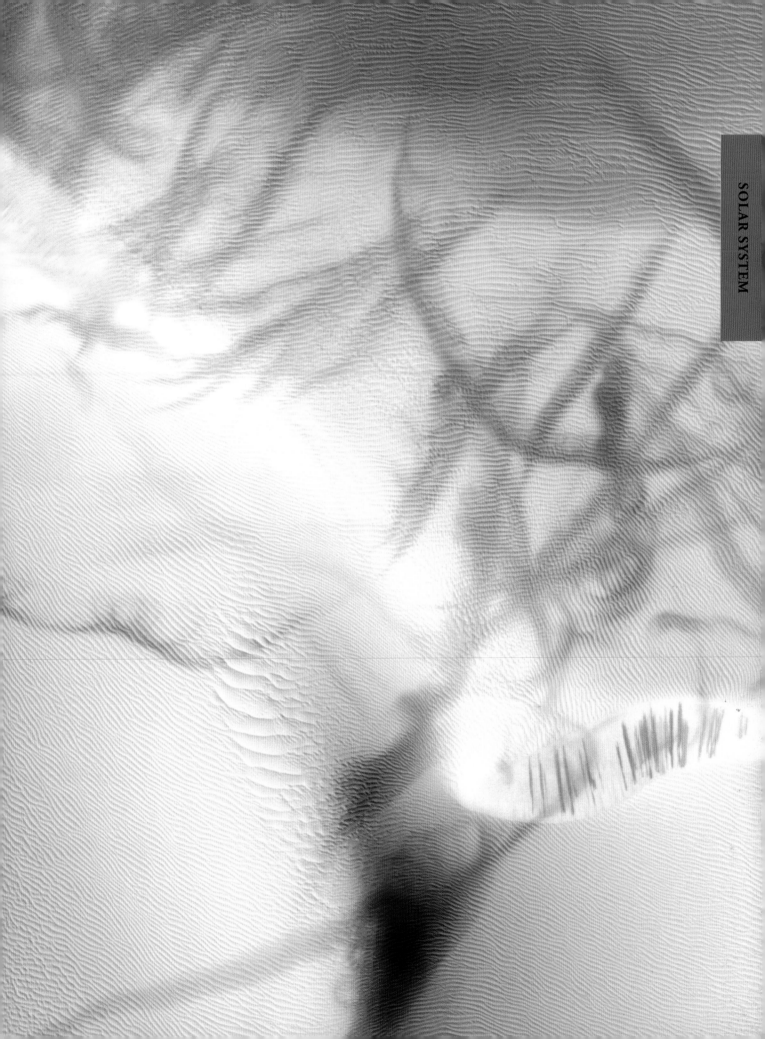

Asteroids

For thousands of years, there were six recognized planets (including Earth) in the solar system. No one dreamt that there were any worlds beyond Saturn, but there were suggestions that something existed between Mars and Jupiter. Rather than a single planet, many thousands of rocky objects have since been discovered. These are asteroids.

📷 WHAT A STAR!

In 1772, Johann Bode proposed a formula to work out the distances of the planets from the Sun. Bode's Law seemed proven by the discovery of Uranus and of Ceres in Bode's "gap" between Mars and Jupiter, but it failed when Neptune and Pluto were discovered.

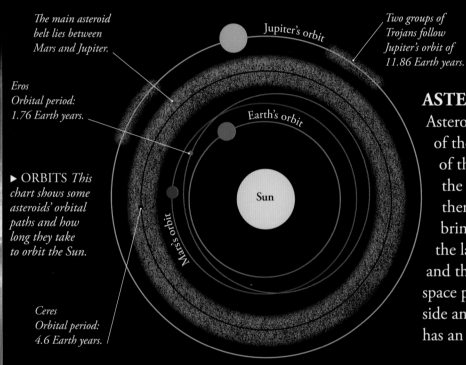

The main asteroid belt lies between Mars and Jupiter.

Jupiter's orbit

Two groups of Trojans follow Jupiter's orbit of 11.86 Earth years.

Eros
Orbital period: 1.76 Earth years.

Earth's orbit

Sun

Mars's orbit

▶ ORBITS *This chart shows some asteroids' orbital paths and how long they take to orbit the Sun.*

Ceres
Orbital period: 4.6 Earth years.

ASTEROIDS IN ORBIT

Asteroids are leftovers from the formation of the planets 4.5 billion years ago. Most of them travel around the Sun between the orbits of Mars and Jupiter, although there are some groups whose orbits bring them close to Earth. Eros is one of the largest of these near-Earth asteroids and the first asteroid to be orbited by a space probe. With a large crater on one side and a depression on the other, Eros has an uneven shape, like a cosmic potato.

Ceres

On January 1, 1801, Giuseppe Piazzi, director of the Palermo Observatory in Sicily, found a mysterious object in the constellation of Taurus. It was found to follow a nearly circular, planetlike path between Mars and Jupiter—but it was too small to be a planet. Today, the object Piazzi named Ceres is classified as a dwarf planet. It is the largest of the asteroids and may have a layer of ice beneath its surface.

Vesta

Vesta is the brightest of the main belt asteroids and is occasionally visible to the naked eye. The asteroid has a giant impact crater 285 miles (460 km) across—nearly as wide as Vesta itself. Vesta was strong enough to survive the huge impact, but some of the debris still falls to Earth as meteorites.

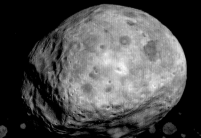

Asteroid tourist

NASA's Dawn spacecraft, launched in 2007, studied both Vesta and Ceres up close.

Hayabusa2

In 2018, a Japanese space probe arrived at the 0.6-mile-wide (1-km-wide) asteroid Ryugu, aiming to study this small world in detail and bring samples of its surface back to Earth.

Ryugu

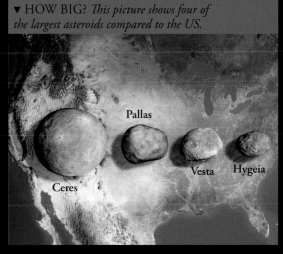

▼ HOW BIG? *This picture shows four of the largest asteroids compared to the US.*

Ceres

Pallas

Vesta

Hygeia

TOO CLOSE FOR COMFORT!

There are far more small asteroids than large ones. Nearly every week, a small asteroid passes close to Earth. There are thought to be 1,100 near-Earth asteroids bigger than 0.6 miles (1 km) across and more than a million longer than 131 ft (40 m). Some have collided with Earth in the past.

CHICXULUB *is a crater in Mexico left by an asteroid that collided with Earth 66 million years ago.*

What's in a name?

The astronomer who discovers a new asteroid has the right to name it. Asteroids are usually named after people, but among the more unusual names are Dizzy, Dodo, Brontosaurus, Humptydumpty, and Wombat.

Crater, fracture, or shatter?

Collisions are common among asteroids, but what happens when they collide depends on how large the asteroid is. If a small asteroid hits a larger one, it will leave a crater. Slightly bigger asteroids may fracture the large asteroid, but the fragments clump back together to form a ball of rubble. If an asteroid is big enough or traveling fast enough, it could shatter a large asteroid, leaving a trail of mini asteroids orbiting in its wake.

▶ TWO WORLDS COLLIDE
When the solar system first formed, asteroids continually collided and grew in size until only one large rocky body was left in an orbit. This became a planet. (◉ pp.120–121)

Jupiter

Jupiter is the king of the planets. This huge world has more than two-and-a-half times the mass of all the other planets combined. Around 1,300 Earths would fit inside this giant world, but because it is mainly made up of light gases, Jupiter weighs only 318 times as much as Earth.

▶ WHAT'S INSIDE?
Jupiter has a relatively small solid core. Most of the planet is made up of hydrogen and helium. Near the surface, the gases are cold, but closer to the core, they get hotter and act more like liquid metal.

Hydrogen and helium gas

Outer layer of liquid hydrogen and helium

Inner layer of metallic hydrogen

Core of rock, metal, and hydrogen compounds

CLOUDS OF MANY COLORS

Ninety percent of Jupiter's atmosphere is hydrogen gas. Most of the rest is helium, with some hydrogen compounds such as methane, ammonia, water, and ethane. The compounds condense (turn to liquid) at different temperatures, making different types and colors of cloud at different altitudes.

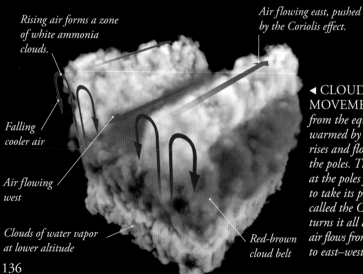

Rising air forms a zone of white ammonia clouds.

Air flowing east, pushed by the Coriolis effect.

Falling cooler air

Air flowing west

Clouds of water vapor at lower altitude

Red-brown cloud belt

◀ CLOUD MOVEMENT *As air from the equator gets warmed by the Sun, it rises and flows toward the poles. The cooler air at the poles flows back to take its place. A force called the Coriolis effect turns it all around so the air flows from north–south to east–west.*

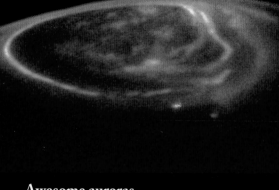

▼ POLAR GLOW *The auroras at Jupiter's poles are hundreds of times more powerful than those on Earth.*

Awesome auroras

Like Earth, Jupiter has a magnetic field, as if there were a giant magnet buried deep inside the planet. It causes auroras (also known as the northern and southern lights). When solar wind particles collide with atmospheric gases, the gases glow and "curtains" of auroral light spread out several hundred miles (kilometers) above Jupiter's clouds.

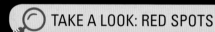

🔍 TAKE A LOOK: RED SPOTS

The most famous feature on Jupiter is the Great Red Spot. This is a giant atmospheric storm that was first recorded in 1831 and has been blowing nonstop ever since. The storm turns counterclockwise once every 6 days. The chemicals that give the Spot its orange-red color are still not known, but the Spot is colder than nearby clouds. In recent years, two more red spots appeared on Jupiter in the same band of clouds.

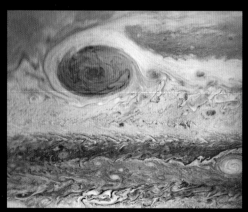

▲ THE HUBBLE *telescope took this image in May 2008. It shows a new red spot to the left of the Great Red Spot and Red Spot Junior.*

Belts and bulges

The white bands of clouds around Jupiter are called zones, and the red-brown bands are belts. Despite its enormous size, Jupiter spins once every 9 hours 55 minutes—faster than any other planet. This makes the clouds at the equator move at more than 28,000 mph (45,000 kph) and causes the equatorial region to bulge outward.

WATCH THIS SPACE

Jupiter is orbited by thin, dark rings of dust. The rings were discovered by Voyager 1 when it flew past the planet in 1979. The main rings are about 155,000 miles (250,000 km) across. The particles in each ring range from microscopic dust to chunks several feet (meters) across.

Over the poles

In 2016, a NASA probe called Juno arrived at Jupiter and entered an unusual orbit that passes over and under the planet's north and south poles. From this angle, Juno has gathered unique information about Jupiter's interior, its powerful magnetism, and the energy source that heats its clouds from beneath.

▲ JUNO'S UNIQUE *view of Jupiter from high above the south pole.*

North polar region

Storm system

North Temperate Zone

North Temperate Belt

North Tropical Zone

North Equatorial Belt

TELL ME MORE …

This picture is made from a set of images taken by the Cassini spacecraft as it traveled 6 million miles (10 million km) away from the planet.

Equatorial Zone

South Equatorial Belt

South Tropical Zone

South Temperate Belt

Great Red Spot

South polar region

PLANET PROFILE

- **Average distance from the Sun** 484 million miles (780 million km)
- **Cloud-top temperature** −234°F (−143°C)
- **Diameter** 89,000 miles (143,000 km)
- **Length of day** 9.93 hours
- **Length of year** 11.86 Earth years
- **Number of moons** about 80
- **Gravity at cloud tops (Earth = 1)** 2.53
- **Size comparison**

Jupiter's moons

Jupiter has about 80 known moons: four "Galilean moons," four inner moons, and the rest are small outer moons. The Galilean moons (Io, Europa, Callisto, and Ganymede) were first discovered in 1610, but very little was known about them until the two Voyager spacecraft imaged them in 1979.

WHAT A STAR!

On January 7, 1610, Italian scientist Galileo Galilei looked through his small telescope and found three small, bright "stars" in a straight line near Jupiter. After weeks of observation, he concluded that there were actually four stars—each a large satellite orbiting the planet. We now call these the Galilean moons.

Io with cheese on top

Io is about the same size as Earth's moon, but it looks like a giant pizza. This is because it's covered by sulfur, which is usually yellow. When sulfur is heated, it changes color, first to red and then to black. The temperature of some of these hot spots can reach 2,730°F (1,500°C). Io is the most volcanically active object in the solar system. There are often a dozen or more volcanoes blasting umbrella-shaped clouds of gas and sulfur compounds into space.

Sulfur dioxide from a volcano settles as a ring of "snow" on the surface.

The black areas scattered over the surface are all active volcanoes.

Plume of gas from the Pele volcano

Pele's plume

Pele is one of Io's largest volcanoes. When Voyager 1 passed it, a plume of gas and dust was rising 190 miles (300 km) above the surface and covered an area the size of Alaska. It can rise high above the moon before falling back to the surface because the gravity on Io is very low. The volcano is surrounded by a blanket of material thrown out during repeated eruptions that has fallen back down to the surface.

The craters of Callisto

Callisto is the most distant of the large Galilean moons. Its surface is billions of years old and is one of the most heavily cratered objects in the solar system. Only a little smaller than Mercury, Callisto is a mixture of ice and rock and has a very weak magnetic field. It also seems to have a salty ocean deep beneath the surface—even though Callisto is not tidally heated like Io, Europa, and Ganymede. Tidal heating happens when the moon is warmed up from the inside, pulled by the gravity of Jupiter and the other Galilean moons.

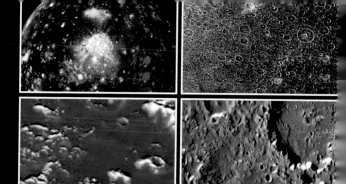

▲ CLOSE-UP CRATERS *These views of Jupiter's second-largest moon reveal that what appear to be lights are actually craters.*

🔍 TAKE A LOOK: EUROPA

Europa is a similar size to Io (and Earth's Moon). It has a smooth surface covered in ice—there are no deep valleys or high mountains and very few impact craters. This shows that its surface is very young. The ice is continually being renewed from below. In fact, parts of the surface look like broken ice floating in the Arctic on Earth. It is thought that Europa has an ocean of water under the outer shell of ice, no more than 6–12 miles (10–20 km) below the airless surface. This is made possible by tidal heating.

◀ 👁 FIND OUT MORE *about the inside of Europa on p.163.*

▲ ICY SURFACE *The white and blue areas in this picture show a layer of ice particles covering Europa's crust. It's thought that the dust came from the creation of a large crater about 620 miles (1,000 km) south of the area.*

Giant Ganymede

With a diameter of 3,270 miles (5,260 km), Ganymede is the largest satellite in the solar system. It is bigger than Mercury but has only about half its mass because Ganymede is a mixture of rock and ice. The interior is thought to be separated into three layers: a small, iron-rich core surrounded by a rocky mantle with an icy shell on top. The surface is divided into two different types of landscape: very old, dark, highly cratered regions; and younger, lighter regions with grooves, ridges, and craters. Ganymede has a weak magnetic field and may have a salty ocean buried 125 miles (200 km) beneath the icy surface.

Did you know
that you can easily see Jupiter from Earth? It appears like a bright star at night and is visible at different times at different locations on Earth. Jupiter is one of the brightest planets—only the Moon and Venus outshine it. You can also see the four Galilean moons with just a small telescope (or a good pair of binoculars, if held steadily).

Ganymede's dark regions are old and full of craters.

Arbela Sulcus is a light region of furrows and ridges 15 miles (24 km) wide surrounded by dark regions.

The lighter regions are younger and have lots of unusual groove patterns.

Voyager 1 & 2

On August 20, 1977, Voyager 2 lifted off from Cape Canaveral, Florida. Voyager 1 followed on September 5. They are two of only four spacecraft to have ever been sent out beyond the solar system. The other two craft, Pioneer 10 and 11, are no longer in touch with Earth, but we still receive regular data transmissions from the Voyagers—even though they have now reached interstellar space.

FAST FACTS

■ Voyager 2 was launched 2 weeks before Voyager 1, but it was on a slower trajectory (path), so Voyager 1 got to Jupiter first.

■ Voyager 1 completed its main mission in November 1980 after a flyby of Saturn's moon Titan.

■ Although their mission was intended to be only a 4-year trip to Jupiter and Saturn, the launch dates allowed Voyager 2 a boost from Saturn, sending it toward Uranus and Neptune.

■ The Voyagers eventually chalked up a wealth of discoveries about all four planets and 48 of their moons.

▶ VOYAGER 1 *was launched into space aboard a Titan III/Centaur rocket.*

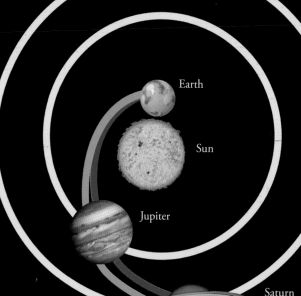

Earth

Sun

Jupiter

Saturn

Uranus

To boldly go
Voyager 1 is the farthest human-made object in space. In August 2012, it crossed the heliopause, the outer boundary of the solar system. Voyager 1 was then 11 billion miles (18 billion km) from the Sun, 120 times the distance of Earth.

SPACE HOPPING

When the Voyagers were launched, Jupiter, Saturn, Uranus, and Neptune were in a rare alignment that only occurs every 175 years. The Voyagers were able to use the powerful gravity of the planets to boost their speed and change direction so they could fly on to the next planet. Voyager 1 arrived at Jupiter in March 1979; Voyager 2 followed in July. Voyager 1 was sent off course by Saturn, but Voyager 2 went on to Uranus and Neptune.

Voyager 1's encounter with Saturn bent the spacecraft's flight path on a course toward interstellar space, preventing it from continuing on to the outer planets.

We've got the power—just!

Each Voyager carried 10 instruments to investigate the planets and their moons. They get their electricity from nuclear power packs. Over time, the power levels have dropped and the output is now about equal to two 120-watt light bulbs. Their computer power is also tiny by modern standards—they both have three computers with 8,000 words of memory each.

Star trek

The Voyagers are leaving the solar system and heading into the Milky Way Galaxy in different directions. Scientists estimate that in about 40,000 years, each spacecraft will be in the neighborhood of other stars and about 2 light-years from the Sun. So far, the Voyagers have traveled just beyond the solar system's outer boundary, known as the heliopause, where the solar wind gives way to interstellar space. Both spacecraft have enough electrical power and attitude-control propellant to continue operating until about 2025.

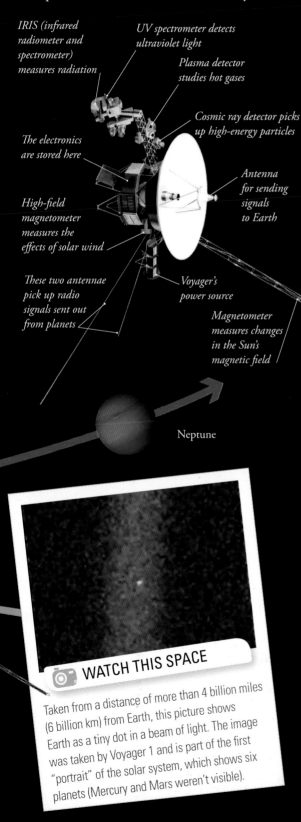

IRIS (infrared radiometer and spectrometer) measures radiation

UV spectrometer detects ultraviolet light

Plasma detector studies hot gases

Cosmic ray detector picks up high-energy particles

The electronics are stored here

Antenna for sending signals to Earth

High-field magnetometer measures the effects of solar wind

These two antennae pick up radio signals sent out from planets

Voyager's power source

Magnetometer measures changes in the Sun's magnetic field

Neptune

TERMINATION SHOCK
Solar wind (a thin stream of electrically charged gas) blows outward from the Sun until it reaches the termination shock. It then drops abruptly as it meets oncoming interstellar wind.

HELIOPAUSE
The heliosphere boundary is where the pressures of the solar wind and the interstellar wind balance. When the Voyagers crossed this boundary, they were in interstellar space.

Voyager 1

Voyager 2

HELIOSHEATH
This is the outer edge of the heliosphere (a huge bubble containing the solar system, solar wind, and the solar magnetic field). Voyager 1 entered the heliosheath about 8.7 billion miles (14 billion km) from the Sun.

BOW SHOCK
As the heliosphere travels through interstellar space, it forms a bow shock, just like waves form around a rock in a stream.

WATCH THIS SPACE

Taken from a distance of more than 4 billion miles (6 billion km) from Earth, this picture shows Earth as a tiny dot in a beam of light. The image was taken by Voyager 1 and is part of the first "portrait" of the solar system, which shows six planets (Mercury and Mars weren't visible).

The Voyager record

Both Voyagers carry a message that will tell any alien life they encounter about where they have come from. The message is carried on a phonograph record—a 12 in (30 cm) gold-plated copper disk containing sounds and images selected to show the variety of life and culture on Earth. The cover shows Earth's location and has instructions on how to play the record. The contents include images, a variety of natural sounds, music from different cultures and ages, and greetings in 55 languages.

Saturn

The second largest planet and sixth planet from the Sun, Saturn is the most distant planet we can see without a telescope. It's visible for about 10 months of the year and is surrounded by an amazing series of rings (but you will need a telescope to see them).

▲ RING CYCLES *Sometimes we see the north side of Saturn's rings and sometimes the south. This is because the orbits of Earth and Saturn are not on the same level, so sometimes Earth is above the rings and sometimes below them.*

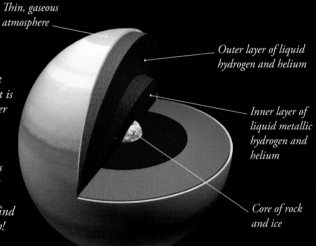

Thin, gaseous atmosphere

Outer layer of liquid hydrogen and helium

Inner layer of liquid metallic hydrogen and helium

Core of rock and ice

▶ LARGE BUT LIGHT *More than 750 Earths could fit inside Saturn, but it is only 95 times heavier than Earth. This is because it is mainly made of hydrogen and helium gas. It is the only planet light enough to float on water—if you can find an ocean big enough!*

TELL ME MORE …

Saturn's rings were first seen by Galileo in 1610, but through his simple telescope, they looked like ears sticking out from the planet!

▲ SATURN ROCKS *Saturn's rings are made up of dust, rocks, and chunks of water ice. They cover a distance of 175,000 miles (280,000 km) but are only about 0.6 miles (1 km) thick.*

RINGS GALORE

Saturn's rings are so spectacular, it is often known as the ringed planet (even though Jupiter, Uranus, and Neptune also have rings). There are three main rings, which are so large and bright that they can be seen in a small telescope. Going outward from the planet, they are known as C, B, and A. Outside these are the F, G, and E rings, which are very faint.

▲ C RING *Inside the C ring is a thin ring called D. There is no gap between these two rings.*

▲ B RING *The widest main ring at 15,850 miles (25,500 km) across. It is 15–50 ft (5–15 m) thick and is also the brightest of the main rings.*

▲ A RING *The first ring to be discovered. The rings are named in order of discovery, not their position.*

◀ MIND THE GAP *Some parts of the rings have been swept clear by the gravity from Saturn's moons, leaving gaps between the rings. The largest gap is the Cassini Division, between the A and B rings.*

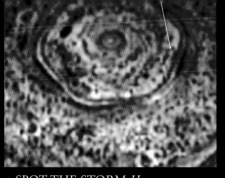

High speed winds spread the storm clouds around the planet's equator. Wind speed at the equator can be 1,100 mph (1,800 kph), six times faster than the strongest winds on Earth. Huge storms also occur at the poles. These have an "eye" like a hurricane. Similar polar storms are found on Venus and Jupiter.

▲ SPOT THE STORM *Huge, hurricanelike storms lie above Saturn's poles. Small storm clouds (shown here as dark spots) move around these huge "whirlpools" in Saturn's atmosphere.*

▲ WATCH THE DRAGON *In Saturn's southern hemisphere is a band of clouds called "storm alley" because so many storms have occurred there—including the large, bright, electrical storm called the Dragon Storm.*

 TAKE A LOOK: POLAR LIGHTS

Saturn's strong magnetic field forms an invisible bubble around the planet. This protects it from most of the electrically charged particles that flow past the planet in the solar wind. However, some of these particles become trapped and flow down the magnetic field lines toward Saturn's magnetic poles. When they strike the upper atmosphere, they form rings of light called auroras.

▶ SOUTHERN LIGHTS *This aurora formed at Saturn's south pole in January 2005.*

PLANET PROFILE

- **Average distance from the Sun**
887 million miles (1,427 million km)
- **Cloud-top temperature** –290°F
(–180°C)
- **Diameter** 74,900 miles
(120,540 km)
- **Length of day** 10.6 hours
- **Length of year** 29.4 Earth years
- **Number of moons** 82
- **Gravity at cloud tops**
(Earth = 1) 1.07
- **Size comparison**

round major moons, smaller irregular inner moons, and tiny irregular outer moons that lie way beyond Saturn's rings. Some of the small moons lie within or very near to Saturn's rings. The outer moons may be comets that were captured by Saturn's powerful gravity. There are also seven "medium-sized" moons that orbit very close to Saturn.

WHAT A STAR!

Dutch astronomer Christiaan Huygens discovered the first of Saturn's moons, Titan, in 1655. The European Space Agency's Saturn probe was named for him.

TELL ME MORE …

Saturn's moons are so cold, their icy surfaces are as hard as rock. They all have impact craters where comets have thumped into them.

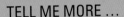

Pan · Atlas · Prometheus · Janus · Enceladus · Pandora · Epimetheus · Mimas · Tethys · Dione · Rhea

▲ **MANY MOONS** *The moons within or close to Saturn's main rings are (from left to right): Pan, Atlas, Prometheus (above), Pandora (below), Janus (above), and Epimetheus (below). Mimas, Enceladus, Tethys, Dione, and Rhea all lie well outside the main rings but within or near to the thin E ring.*

▶ PHOEBE *Like most of Saturn's moons, the outer moon called Phoebe travels in an elliptical (oval-shaped) orbit. It has created its own ring of ice and dust, known as Phoebe's Ring, 3.7–7.4 million miles (6–12 million km) outside Saturn.*

▲ **HYPERION** *Most of Saturn's moons keep the same face toward the planet. However, Hyperion tumbles over as it orbits Saturn. This may be due to one or more collisions with comets.*

◀ IAPETUS *The 24th moon from the planet, Iapetus is Saturn's most distant major moon. Its forward-facing side is covered in dust that has been knocked off Phoebe by comet collisions. Unlike nearly all the other moons, it travels in the same direction as Saturn.*

▲ TITAN *The second largest moon in the solar system (Jupiter's Ganymede is the first), Titan is bigger than the planet Mercury. Its orbit is 758,000 miles (1.2 million km) from Saturn.*

Saturn's largest moon is unique—it is the only moon to have an atmosphere. Titan's atmosphere is nitrogen-rich and dense like Earth's, but it is far too cold to support life. Radar and infrared instruments have been used to study Titan's surface, which is hidden beneath a thick orange haze. The surface was found to be covered in ice, with mountains, huge dunes, and rivers and lakes of liquid methane.

▲ CASSINI-HUYGENS
The Cassini orbiter spent over 13 years studying Saturn and its main moons. The Huygens probe was designed to explore Titan's atmosphere and surface.

▶ TITAN'S *surface has channels that were probably carved out by flowing methane. On Earth, methane is a gas, but Titan is so cold (–290°F/–179°C), methane is a liquid and falls as rain from the clouds.*

◀ POLES APART
Taken 2 months later, in December 2005, this view is of Titan's opposite hemisphere (the "back" of the first image). You can clearly see the north and south polar regions.

▲ BRIGHT LIGHTS *This false-color image was taken by the Cassini spacecraft. The very bright area is called Tui Regio and is thought to be frozen water or carbon dioxide that has come from a volcano.*

🔍 TAKE A LOOK: ENCELADUS

Perhaps the most surprising of Saturn's moons is Enceladus. Only about 300 miles (500 km) across, Enceladus was expected to be a cold, dead world. However, the Cassini spacecraft discovered powerful geysers near the moon's south pole. Tidal movement inside the moon creates heat that turns ice into water vapor. This escapes through cracks, or fault lines, in Enceladus's icy shell and is blasted into space.

Water ice particles in the geysers feed the E ring that circles Saturn.

The water becomes much warmer near the surface.

▲ IT'S YOUR FAULT *The plumes of gas and icy particles blast into space through large fault lines in the surface known as "tiger stripe" fractures.*

Touchdown on Titan

After a 2.5 billion mile (4 billion km) piggyback ride lasting almost 7 years, the European Space Agency's Huygens probe separated from the Cassini orbiter on December 25, 2004. It landed on Titan on January 14, 2005, making it the first (and, so far, only) time that a spacecraft had touched down on a world in the outer solar system. The probe's instruments swung into action, sampling the atmosphere and taking pictures.

(1) The top image shows the area where Huygens landed. (2) View as Huygens parachutes to Titan, taken from 4 miles (6 km) above. (3) Titan's tallest mountains are thought to be just a few hundred yards (meters) tall. (4) Artist's impression of Huygens.

SATURN BY SUNLIGHT

This amazing view of Saturn directly in line
with the Sun is made up of 165 images taken
by the Cassini orbiter. Lit from behind, the
planet is in shadow, but the glow reveals
previously unseen, unknown rings—and,
billions of miles in the distance, Earth.

Earth

Uranus

Uranus is the third largest planet and the seventh planet from the Sun. At this distance, it receives little heat or light from the Sun, so the cloud tops are extremely cold. Each orbit around the Sun takes 84 Earth years, so birthdays on Uranus are extremely rare!

GAS AND ICE

Around 63 Earths would fit inside Uranus, but since it is mostly made of gas, it is only 14 times heavier than Earth. Uranus and Neptune are sometimes called ice giants because a large part of their interiors is thought to be composed of ices made from water, methane, and ammonia.

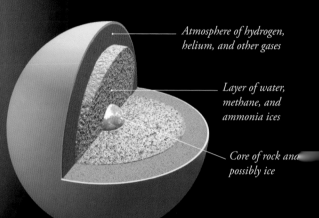

Atmosphere of hydrogen, helium, and other gases

Layer of water, methane, and ammonia ices

Core of rock and possibly ice

📷 WHAT A STAR!

William Herschel discovered Uranus in 1781. Looking through his homemade telescope, he noticed a greenish star in the constellation of Gemini that was not shown on his sky charts. Herschel thought it was a comet, but a year later, it was confirmed as a new planet.

🔍 TAKE A LOOK: BLACK RINGS

Uranus has a system of 13 dark, thin rings around the planet. They are very black, extremely narrow—less than 6 miles (10 km) across—and mostly made of dust and boulders up to 3 ft (1 m) across. The rings are too faint to be seen from Earth and were only discovered in 1977, when the planet passed in front of a star. The light from the star was dimmed as it passed through the rings.

Clouds on Uranus

Through most of Earth's largest telescopes, Uranus appears as an almost featureless disk. When Voyager 2 flew past the planet in 1986, it sent back images of a pale blue ball with a few clouds or storm features. The Hubble Space Telescope has since found that some large clouds travel around the planet more than twice as fast as hurricane winds on Earth.

▲ KECK'S CLOUDS *This picture, taken in infrared light by the Keck Telescope in Hawaii, shows the planet's rings and storm clouds, which are harder to spot in visible light.*

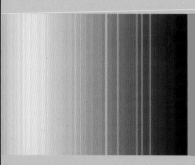

▲ LINE UP *The outermost ring, Epsilon, is shown as a white line in this false-color image.*

Uranus's moons

Uranus has a family of 27 known moons, many of them named after characters from Shakespeare's plays. Most of these are small objects less than 125 miles (200 km) across that orbit the planet close to the rings. They include Cordelia and Ophelia, which are "shepherd moons"—they keep the particles of the thin Epsilon ring in place.

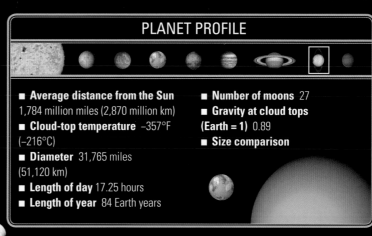

PLANET PROFILE

- **Average distance from the Sun** 1,784 million miles (2,870 million km)
- **Cloud-top temperature** –357°F (–216°C)
- **Diameter** 31,765 miles (51,120 km)
- **Length of day** 17.25 hours
- **Length of year** 84 Earth years
- **Number of moons** 27
- **Gravity at cloud tops (Earth = 1)** 0.89
- **Size comparison**

Oberon

Titania

Umbriel

Uranus

Miranda

Ariel

Major moons

The five major moons of Uranus are cold, icy worlds. Miranda is the smallest. Ariel is the brightest and was discovered in 1851 at the same time as the heavily cratered Umbriel. Titania and Oberon, the two largest moons, show some signs of internal warming in the past.

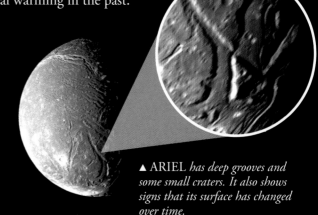

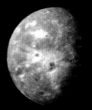

▲ OBERON *was the first moon to be discovered, by William Herschel in 1787.*

▲ ARIEL *has deep grooves and some small craters. It also shows signs that its surface has changed over time.*

Miranda

Miranda has unique surface features, including deep canyons, terraced layers, and much younger, smoother layers. These point to a turbulent history. Some suggest that Miranda suffered a catastrophic collision in the distant past and then reassembled in the chaotic way that we see today. Alternatively, it may have started to evolve, with heavier material sinking toward the center and lighter material rising to the surface, but this process stopped before it was completed.

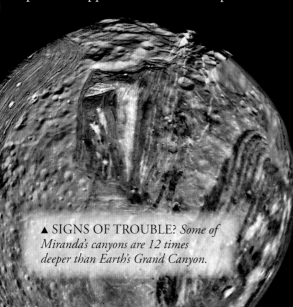

▲ SIGNS OF TROUBLE? *Some of Miranda's canyons are 12 times deeper than Earth's Grand Canyon.*

The toppled planet

Uranus is unusual because it is tipped over on its side so that the equator is almost at right angles to the orbit and its poles take turns in pointing toward the Sun. Each pole has 21 years of permanent sunlight during its summer and 21 years of permanent darkness in its winter. It is believed that Uranus may have been knocked over by a huge collision with a planet-sized body early in its history.

▶ UPRIGHT ORBIT *This Hubble Space Telescope view shows how Uranus's moons follow the tilt of the planet and orbit it top to bottom.*

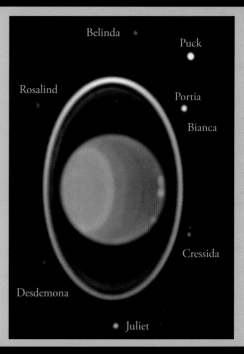

Belinda

Puck

Rosalind

Portia

Bianca

Desdemona

Cressida

Juliet

Neptune

The eighth planet from the Sun, Neptune is an icy gas giant 58 times the size of Earth but only 17 times heavier. It is an extremely cold, dark world—30 times farther from the Sun than Earth, it receives 900 times less light and heat than Earth.

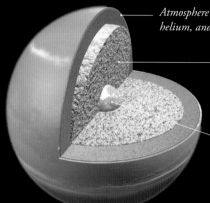

▼ FULL OF GAS *Although it is 58 times the size of Earth, Neptune is mainly made of gas, water, and ices, which makes it relatively light.*

Atmosphere of hydrogen, helium, and methane gases

Icy layer of frozen water, methane, and ammonia

Solid core of rock and possibly ice

A BLUE PLANET

Like Uranus, Neptune appears blue—not because it is covered with oceans, but because it has methane gas in its atmosphere. This gas absorbs red light from the Sun, and when red light is taken away from visible light, it leaves behind blue light.

Active atmosphere

Heat rising from inside Neptune makes the planet's atmosphere very active—it feeds some large storms and drives the fastest winds in the solar system. Cloud features on Neptune have been seen to sweep around the planet at around 1,240 mph (2,000 kph), 10 times the speed of hurricane-force winds on Earth. Sometimes these winds are revealed by long banks of high-level clouds.

◄ SHADOWS *Methane ice clouds cast shadows on the main deck of blue clouds 30 miles (50 km) below. The cloud streaks are 30–125 miles (50–200 km) wide but stretch for thousands of miles around the planet.*

TELL ME MORE ...

Almost everything we know about Neptune comes from the Voyager 2 spacecraft, which flew past the planet in 1989. Neptune was the fourth and last planet visited by Voyager 2 as it headed out of the solar system toward interstellar space.

THE GREAT DARK SPOT

Neptune's atmosphere changes quite quickly as large storms and cloud features rush around the planet in the opposite direction to its rotation. A white cloud feature called Scooter took just 16.8 days to zip around the planet. The largest feature seen so far was the Great Dark Spot, a storm about the same size as Earth. It disappeared within a few years.

NEPTUNE'S MOONS

■ **Neptune has 14 known moons.** The largest of these is Triton, which is smaller than Earth's moon but larger than the dwarf planet Pluto. It travels the opposite way around the planet compared to most other moons and is gradually being pulled toward Neptune. Triton is one of the coldest worlds we know, with a surface temperature of –391°F (–235°C). It is covered by frozen nitrogen gas. Despite the extreme cold, Triton seems to be warm inside.

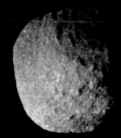

◄ SMALL BUT SPEEDY *Proteus is the largest of the seven inner moons. It takes 27 hours to travel around Neptune.*

◄ TRITON'S TRAILS *Dark trails across Triton's surface show where ice "geysers" throw black dust into the thin atmosphere. This is blown away from the polar region and coats the surface.*

■ **Most of Neptune's outer moons** are small—Nereid is 210 miles (340 km) across, and the others are less than 125 miles (200 km) across. Seven of them orbit close to the planet, within 74,500 miles (120,000 km). Five follow distant orbits more than 9 million miles (15 million km) away and are probably captured comets.

Neptune's rings

Neptune has a system of six very narrow, dark rings. Four small moons lie inside the ring system. Two of these—Galatea and Despina—act as shepherds for the ring particles, keeping two of the rings in shape. Galatea is probably also the cause of the Adams ring being unusually clumpy. This ring has arcs, meaning it is thicker in some places than others.

Orbit oddity
Neptune is normally the eighth planet from the Sun, but it has such an elliptical (oval-shaped) orbit that for about 20 years of its 164-year-long trip around the Sun, it is actually farther away than Pluto. This was the case from 1979 to 1999.

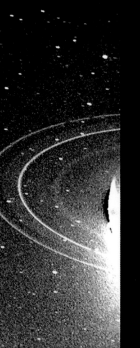

◄ RINGS *These photos from Voyager 2 show four rings. The two bright rings are Adams ring (outer) and Le Verrier (inner).*

Johann Galle

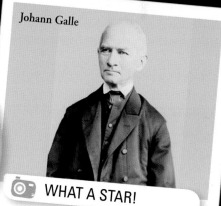

📷 **WHAT A STAR!**

The search for Neptune began when astronomers noticed that something seemed to pull on Uranus so that it sometimes traveled faster than expected and sometimes slower. The new planet was found by Johann Galle in 1846, after its position was worked out by John Couch Adams and Urbain Le Verrier.

PLANET PROFILE

■ **Average distance from the Sun** 2,800 million miles (4,500 million km)
■ **Cloud-top temperature** –364°F (–220°C)
■ **Diameter** 30,760 miles (49,500 km)
■ **Length of day** 16 hours
■ **Length of year** 165 Earth years

■ **Number of moons** 14
■ **Gravity at cloud tops (Earth = 1)** 1.13
■ **Size comparison**

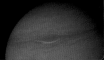

Pluto and beyond

Cold, icy, and far from the Sun, Pluto is the largest of several dwarf planets in our solar system. It is one of the closest and brightest objects in the Kuiper Belt, the ring of icy worlds that surround the solar system beyond the orbit of Neptune.

AN ECCENTRIC ORBIT

Pluto's orbit is tilted at 17 degrees from the plane in which the major planets orbit and is also more elongated than any other orbit, ranging between 2.8 and 4.6 billion miles (4.4 and 7.4 billion km) from the Sun.

▼ PLUTO'S LONG JOURNEY
Pluto takes 248 years to orbit the Sun, on an eccentric path that takes it into the Kuiper Belt and within Neptune's orbit.

Pluto   Kuiper Belt   Neptune   Saturn   Mars   Jupiter   Uranus

▶ ULTIMA THULE
On January 1, 2019, the New Horizons spacecraft flew past this small Kuiper Belt object made up of two fragments that stuck together after a slow collision. It is by far the most remote object targeted by a spacecraft.

THE KUIPER BELT

The Kuiper Belt consists of many millions of objects that orbit the Sun at distances ranging from the orbit of Neptune to about three times farther from the Sun. Kuiper Belt objects are icy debris left over from the formation of the planets 4.5 billion years ago. Orbits in the inner half of the belt are stable, but objects farther out can sometimes have their orbits disrupted, falling toward the Sun, where they become short-period comets.

OTHER DWARF PLANETS

Astronomers define a dwarf planet as a small world that looks like a planet but shares its orbit with many other objects. Aside from Pluto, there are four other known dwarf planets—Ceres in the asteroid belt and Eris, Makemake, and Haumea in the Kuiper Belt. Eris is almost as large as Pluto, has one known moon, and follows a very elliptical 560-year orbit. Haumea is smaller than Pluto, shaped like an airship, and spins on its axis every four hours, while Makemake is even smaller and has a distinctive red color.

Frozen surface

Pluto is a deep-frozen world, with temperatures barely rising above −382°F (−230°C), even at its closest approach to the Sun. Its atmosphere is far thinner than Earth's, with the most common gases being nitrogen and methane. As Pluto orbits the Sun, the gases move back and forth between gaseous and frozen states, leading to big changes in the size and thickness of the overall atmosphere.

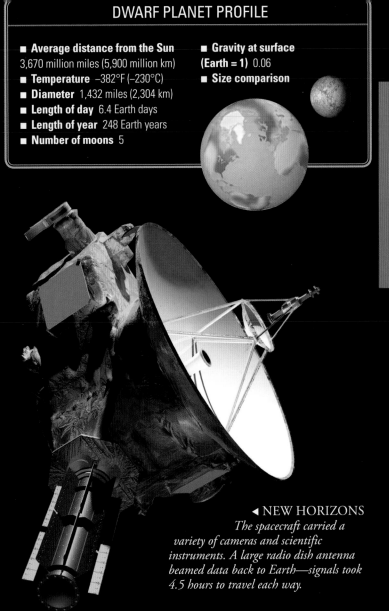

AMAZING ICE WORLD

Pluto is a surprisingly active world. A bright heart-shaped area called Tombaugh Regio, formed by warmer ice pushing up from beneath the frozen crust, is covered with smooth, craterless plains. Elsewhere, there are huge water-ice mountains and ancient cratered terrain. Older, reddish regions are covered in complex chemicals called tholins, which are possible building blocks of life.

Pluto

DWARF PLANET PROFILE

- **Average distance from the Sun** 3,670 million miles (5,900 million km)
- **Temperature** –382°F (–230°C)
- **Diameter** 1,432 miles (2,304 km)
- **Length of day** 6.4 Earth days
- **Length of year** 248 Earth years
- **Number of moons** 5
- **Gravity at surface (Earth = 1)** 0.06
- **Size comparison**

▶ CHARON
Pluto's largest moon, Charon, is about half the size of Pluto and orbits it in 6.4 days. Charon may have started out as a separate dwarf planet before being captured by Pluto's weak gravity.

Charon

THE VIEW FROM EARTH

Seen from Earth, Pluto is nothing more than a very faint star, even to the powerful Hubble Space Telescope. Hubble captured this earthbound view of Pluto in 2005, revealing three of its five known moons.

◀ NEW HORIZONS
The spacecraft carried a variety of cameras and scientific instruments. A large radio dish antenna beamed data back to Earth—signals took 4.5 hours to travel each way.

RACING TO PLUTO

Most of what we know about Pluto is due to a single mission. Launched in 2006, NASA's New Horizons was the fastest spacecraft ever to leave Earth's gravity, traveling at a speed of 10.1 miles (16.26 km) per second. After hurtling across the solar system, it flew past Pluto at a distance of 7,800 miles (12,500 km) in July 2015.

Nix

Pluto

Hydra

Charon

Daytime darkness
If people lived on Pluto, they would need flashlights. Even in the daytime, light levels are between 900 and 2,500 times lower than on Earth.

Comets

Every now and then, a strange object with a wispy tail appears in the night sky. This is a comet, a large lump of dust and ice a few miles long hurtling toward the Sun. There are billions of comets circling the Sun, far beyond the orbit of Pluto.

Collision course
Sometimes a comet can be nudged out of its orbit so that it travels into the inner solar system. If it hits Earth, it may result in widespread destruction. But don't worry—the chance of this happening is very small!

DIRTY SNOWBALLS

The nucleus (solid center) of a comet is made of dirty water ice. The "dirt" is rock dust. When a comet warms up, the nucleus releases gas and dust. They form a cloud called a coma. Sometimes, long tails develop and extend millions of miles into space. There are two main tails: a bluish gas tail and a white dust tail. The tails always point away from the Sun.

Nucleus made of water ice and silicate rock dust

Black crust made of carbon

Bright side faces the Sun

Dust tail is curved

Gas tail

Perihelion

Tails are longest close to the Sun

Sun

Tail grows as comet moves toward the Sun

Naked nucleus

Aphelion (farthest point from Sun)

Jets of gas and dust

LIFE CYCLES

A comet spends most of its life in a frozen state until it moves near the Sun, when it warms up and gets active. The coma is largest at the perihelion (the point nearest the Sun), when the icy nucleus is releasing most gas and dust. Each time a comet passes near the Sun, it gets slightly smaller. If a comet stayed on the same orbit for thousands of years, it could eventually evaporate to nothing.

Comet Hale-Bopp
Many new comets are found each year, but few of them can be seen without large telescopes. Sometimes a very bright comet appears in our skies. The great comet of 1997 was comet Hale-Bopp, named after its discoverers, Alan Hale and Tom Bopp. Hundreds of millions of people were able to see the comet after dark with the naked eye.

Halley's comet

Halley is the most famous of all the comets. It is named after Edmond Halley, who first realized that the comets seen in 1531, 1607, and 1682 were actually the same object. Halley worked out that it reappeared every 76 years after traveling out beyond the orbit of Neptune. He predicted that it would return in 1758–1759 and it did, although he did not live to see it. Like many comets, it orbits the Sun in the opposite direction to the planets.

▶ BAD OMEN
Halley features in the Bayeux Tapestry. It appeared just before the Battle of Hastings in 1066.

Fan-tastic tails

Some comets produce spectacular tails that spread out like fans. Comet McNaught, which was the brightest comet for more than 40 years, provided a great example of this in the skies above the southern hemisphere in early 2007. Outbursts of dust created a broad, fan-shaped tail that was visible even in daylight. It was mistaken for a brush fire, an explosion, and a mysterious cloud.

SHATTERING COMETS

A comet's nucleus is not very strong, and sometimes it breaks into small pieces. Comet Shoemaker-Levy 9 was broken into 21 pieces by Jupiter's gravity in 1994. Fragments crashed into the planet, leaving dark spots in the clouds. Other comets have broken up on their orbit near the Sun. In 1995, comet Schwassmann-Wachmann 3 broke into five large pieces. It continues to spilt into smaller and smaller pieces and is likely to soon disintegrate completely.

Jupiter's surface is scarred by comet pieces.

▶ MANY PIECES *form as Shoemaker-Levy breaks up.*

Oort Cloud

Billions of comets are thought to exist in the Oort Cloud, named after scientist Jan Oort. This vast, ball-shaped cloud exists far beyond Pluto, more than 1 light-year from the Sun. The comets spend most of their lives here in deep freeze. Occasionally, when one is disturbed by a passing star, it begins to travel inward toward the Sun. We only know of its existence when it starts to evaporate and grows tails during the approach. Comet Hyakutake, one of the brightest comets of the late 20th century, came from the Oort Cloud. It won't return to Earth's skies for over 100,000 years.

Professor Jan H. Oort

▶ STAR STRUCK
A star passes close to the Oort Cloud and knocks a comet into a new orbit.

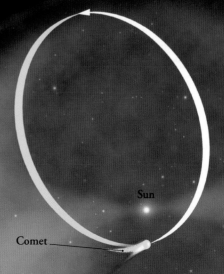

Sun

Comet

Comet missions

Comets were once mysterious visitors to the solar system. Since 1986, we have discovered more about them by sending spacecraft to have a closer look. Probes have not only flown past comets but have also collected samples of comet dust and even crashed into a comet's nucleus.

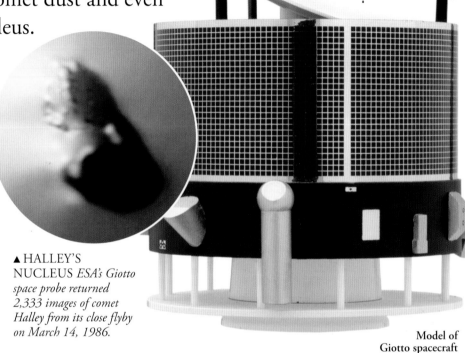

Backup antenna

Dish-shaped main antenna

Model of Giotto spacecraft

Giotto

The first close-up views of a comet's nucleus came from the European Space Agency's Giotto spacecraft. In 1986, it flew past the nucleus of comet Halley at a distance of less than 373 miles (600 km). Images showed a black, potato-shaped object with jets of gas and dust spewing into space from the Sun-facing side. Giotto was damaged by a high-speed impact with a large dust grain but recovered to become the first spacecraft to visit two comets—in 1992, it passed within 124 miles (200 km) of comet Grigg-Skjellerup.

▲ HALLEY'S NUCLEUS *ESA's Giotto space probe returned 2,333 images of comet Halley from its close flyby on March 14, 1986.*

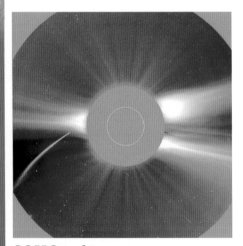

SOHO and its sungrazers

Designed to observe the Sun, the ESA–NASA SOHO spacecraft is able to block out the Sun's glare. This has revealed many "sungrazers"—comets that pass close to the Sun (and usually fall into it). SOHO has discovered nearly 1,700 comets since 1996.

STARDUST

■ NASA's Stardust spacecraft was launched toward comet Wild 2 in February 1999. Stardust was designed to collect dust samples from the comet. The particles were captured in aerogel and brought back to Earth for analysis.

■ In January 2004, Stardust swept past Wild 2 at a distance of 147 miles (236 km). Images taken by the spacecraft revealed the comet to be surprisingly different to comets Borrelly and Halley. Although its hamburger-shaped nucleus was only 3 miles (5 km) across, its surface was strong enough to support cliffs and pinnacles over 328 ft (100 m) high. Most noticeable of all were large circular craters up to 1 mile (1.6 km) wide and 500 ft (150 m) deep.

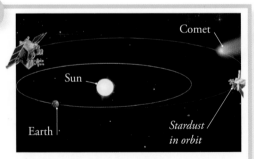

▲ IN SPACE *This artist's impression shows Stardust on its mission to comet Wild 2. It is now on a mission to fly by comet Tempel 1.*

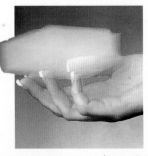

◀ LIGHT AS AIR *Made of 99.8 percent air, the ghostly looking aerogel is the only substance that can collect high-speed comet particles without damaging them.*

Deep Space 1

NASA's Deep Space 1 was launched in October 1998. It passed within 1,400 miles (2,200 km) of comet Borrelly in September 2001, and sent back the best pictures of a nucleus ever seen before. The nucleus measured about 5 miles (8 km) long and 2.5 miles (4 km) wide. It was found to be the blackest object in the solar system, reflecting less than 3 percent of the sunlight that it receives.

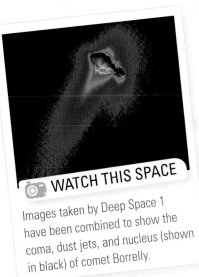

📷 **WATCH THIS SPACE**

Images taken by Deep Space 1 have been combined to show the coma, dust jets, and nucleus (shown in black) of comet Borrelly.

▲ OUTBURST *This Hubble Space Telescope image shows an outburst of ice particles from comet Tempel 1.*

Point of impact

Comet Tempel 1

NEAR and far

NEAR (Near Earth Asteroid Rendezvous) Shoemaker made history when it became the first spacecraft not only to orbit but also land on an asteroid. It touched down on Eros on February 12, 2001, and sent data and images back to Earth. NEAR stopped working on February 28 and remains on Eros.

NEAR Shoemaker is 9 ft 2 in (2.8 m) tall to the top of its antenna.

Deep Impact

To find out more about what a comet is made from, NASA sent its Deep Impact mission to collide with comet Tempel 1. The probe released by the spaceship collided with the nucleus at 22,370 mph (36,000 kph) and exploded on arrival, throwing out a huge cloud of ice and dust and creating a stadium-sized crater. The nucleus was revealed to be 3.1 miles (5 km) long and 4.3 miles (7 km) wide, with ridges and curved slopes.

Comet 67P

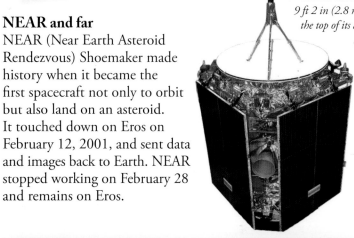

▲ THE PHILAE *lander lost contact with Rosetta after landing in a shadowy crack on the comet's surface.*

Rosetta

The European Space Agency's Rosetta probe has been the most revealing comet mission yet. Launched from Earth in 2004, it took more than 10 years to reach its target, Comet 67P/ Churyumov-Gerasimenko. The probe then entered orbit for 2 years, using a dozen science instruments to study every aspect of Comet 67P as it made its closest approach to the Sun and jets of gas and dust began to erupt through its crust. Finally, in September 2016, Rosetta made a slow descent to settle forever on the icy surface.

Meteors

Look up into the night sky and you might just see a brief trail of light left by a meteor. Also called shooting stars, meteors appear without warning and usually last less than a second. They are particles of dust that burn up as they hit the upper atmosphere at high speed—25,000 mph (40,000 kph) or more.

A METEOR SHOWER

The best time to look out for meteors is during an annual shower. They appear around the same dates each year, when Earth passes through a stream of dust left behind by a passing comet. It may be particularly impressive if the comet has passed through the inner solar system quite recently.

TAKE A LOOK: THE CHELYABINSK METEOR

In February 2013, a large meteor about 66 ft (20 m) across streaked through the skies over the Russian city of Chelyabinsk. This giant fireball, now known as the Chelyabinsk meteor, briefly outshone the Sun before exploding at about 14 miles (23 km) above the ground. The resulting shock wave shook buildings and shattered windows. Meteors like this one, called "superbolides," happen only a few times in a century, but most explode unnoticed over the oceans. Even though they are not big enough to reach Earth's surface, the "air bursts" they create can cause devastation if they occur over land.

▲ REGULAR VISITOR *The Perseids originate from a cloud of debris left behind by the comet Swift-Tuttle, as seen in this false-color image. This meteor shower is visible from Earth every year in August.*

The Leonids

First reported by Chinese astronomers in 902 CE, the Leonids appear to come from the constellation Leo. This shower can be seen every year in mid-November, when 10 to 15 meteors per hour are usually visible around peak times. Every 33 years or so, the Leonids go through a period of great activity in which thousands of meteors an hour hurtle across the sky. Although most meteoroids are no larger than a grain of sand, the shower can be so active that it looks like falling snow.

— Star trail

— Meteor

▲ LOTS OF LEONIDS *This Leonid shower occurred over Korea in November 2001.*

Fireballs

Extremely bright meteors are known as fireballs. They occur when a small piece of rock becomes very hot and bright as it enters Earth's atmosphere. Some fireballs are so bright that they are visible in daylight, and some can create a very loud sonic boom (like an aircraft breaking the sound barrier) that can shake houses. Sometimes the chunks of rock explode, scattering small meteorites on the ground.

▲ FAST FLIGHT *This Leonid fireball moved at a speed of 43 miles (70 km) per second.*

Meteorites

Every year, around 22,000 tons (20,000 tonnes) of cosmic dust and rock enter Earth's atmosphere as meteors. Those that are large enough to survive the fiery entry and reach the ground are called meteorites. Most meteorites that fall to Earth are pieces that have broken off asteroids during collisions in space.

MAKING AN IMPACT

When a meteorite or asteroid lands, it can make a crater.

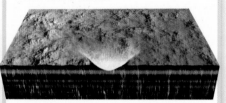

▲ A METEORITE *hits the ground at speed, creating heat that vaporizes it.*

▲ ENERGY *from the impact throws rocks up and out from the ground.*

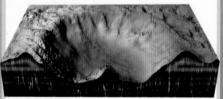

▲ LARGE IMPACTS *cause the crust to rebound, creating a central peak.*

▲ WHO IS HOBA? *Most meteorites are named after the place they fell. Hoba is named for Hoba Farm near Grootfontein, Namibia.*

HEFTY HOBA

The Hoba meteorite is the largest on Earth. The iron meteorite is thought to have landed less than 80,000 years ago and still lies at Hoba Farm, Namibia, where it was found in 1920. Surprisingly, the 132,280 lb (60,000 kg) meteorite did not dig out a crater when it hit the ground, perhaps because it entered the atmosphere at a shallow angle and was slowed down by atmospheric drag.

What's what?
- *Meteoroid* A small piece from an asteroid or comet orbiting the Sun.
- *Meteor* A meteoroid that has entered Earth's atmosphere and burns up brightly.
- *Meteorite* A meteoroid that lands on the Earth's surface.

Meteor Crater
Nearly 200 impact craters have been found on Earth. One of the youngest is in Arizona. Meteor Crater (also called Barringer Crater) was probably excavated about 50,000 years ago by a 300,000 ton (270,000 tonne) iron meteorite. The crater is 3,940 ft (1,200 m) wide, 600 ft (183 m) deep, and surrounded by a wall of loose rock up to 148 ft (45 m) high.

TAKE A LOOK: METEORITE TYPES

Meteorites help us understand conditions in the early solar system 4.5 billion years ago. There are three main types. Stony meteorites are common but tend to break up as they fall to Earth. Iron meteorites are less common in space, but they are very strong and usually land in one piece. "Stony-irons" are a mixture of the two types. Meteorites are usually coated with a black crust that forms when they are heated during passage through the atmosphere.

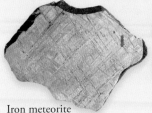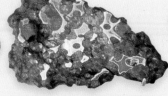

Iron meteorite   Stony-iron meteorite   Stony meteorite

Tagish Lake meteorite

This rare meteorite fell to Earth on the frozen surface of Tagish Lake, Canada, in 2000. The fragile, charcoal-like meteorite is rich in carbon and contains some of the oldest solar system material yet studied.

Meteorites *on* Mars …

Meteorites fall on other worlds, as well as Earth. NASA's Opportunity rover has come across several meteorites on the surface of Mars. The largest of these rocks was found in the Meridiani Planum region in July 2009. Named Block Island, it is made of iron and nickel and may have been lying on Mars for millions of years.

▶ BIG BLOCK
Block Island is 2 ft (60 cm) long and 1 ft (30 cm) wide.

… and meteorites *from* Mars

Of the more than 60,000 meteorites that have been found on Earth, around 250 have been identified as coming from Mars. These rocks were blasted into space long ago by large impacts and traveled through space for many thousands or even millions of years until they fell to Earth. Although no one saw them land, we know that they come from Mars because they contain gases that are exactly the same as those found there. There are also more than 350 named meteorites that have been identified as lunar (coming from the Moon).

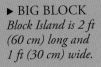

NWA 2626 meteorite

▲ CLOSE-UP CRYSTALS *Found in Algeria in November 2004, the NWA 2626 meteorite comes from Mars. It contains large crystals and glassy veins.*

ANYONE FOR TENNIS? *More than 2,000 tennis courts can fit inside Meteor Crater!*

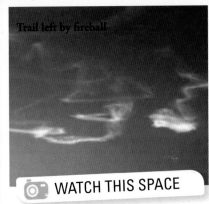

Trail left by fireball

📷 WATCH THIS SPACE

Meteoroid 2008 TC3 became the first to be seen *before* it hit Earth. Spotted out in space, astronomers correctly predicted when and where it would enter Earth's atmosphere: October 7, 2008, in Sudan.

Life on other worlds

Life is found in some surprising places on Earth, from inside solid rock to volcanic vents and the frozen Antarctic. Some experts think that simple organisms may exist in other parts of the solar system—if the ingredients for life can be found.

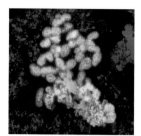

◄ INGREDIENTS FOR LIFE *Life seems to be able to survive where water and a source of energy are present. This slime mold lives on and eats rock.*

LIFE ON MARS?

■ **People have been fascinated** by the possibility of life on Mars for centuries. They assumed that Mars's orbit just beyond Earth might give it Earthlike conditions, and some even reported seeing artificial canals running across the surface, but these were later proved to be an optical illusion. Space probes have shown that Mars is actually a dry, cold desert, but they've also shown that conditions could have been suitable for life in the distant past.

Criss-crossing canals

Fantastical and outdated map of Mars

► MICROLIFE *Inside the Martian meteorite were tiny wormlike structures and magnetite crystals, which are associated with some kinds of bacteria.*

■ **In the 1990s**, NASA scientists reported finding evidence of ancient life in a Martian meteorite—a chunk of rock that was blasted off Mars by an impact and later fell to Earth in Antarctica. They found chemicals and structures that, on Earth, are normally only formed by the activity of bacteria, and even potential "fossils." However, it will take much more evidence to conclusively prove the case for ancient Martian life.

■ **As NASA Mars rovers** such as Opportunity and Curiosity have explored the Martian surface rocks over the past two decades, they've found evidence that Mars was once warmer, wetter, and more Earthlike, including signs of ancient hot springs similar to those where life on Earth may have first evolved.

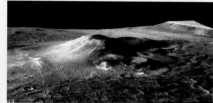

▲ BRIGHT SILICA DEPOSITS *discovered on the flanks of an extinct Martian volcano may have been formed by hot springs.*

Life in the clouds

The gas giant Jupiter has no solid surface or oceans of water, but scientists have suggested that life forms could exist floating in the clouds. Such life could only survive in the upper atmosphere, as the pressure and temperature are too high in the lower atmosphere. However, probes have found no evidence of life at all.

▲ SWIMMING IN THE SKY *Could alien life forms on gas giants behave like jellyfish or rays floating in Earth's oceans?*

EUROPA

Scientists think that Jupiter's ice-covered moon Europa is the most likely location for extraterrestrial life in the solar system. Europa's surface is covered with fractured ice, but below the surface may be a hidden ocean where life might flourish. There might even be hot, hydrothermal vents on the sea floor. On Earth, such vents are surrounded by strange life forms and are considered a likely site for the origin of life on our planet.

▲ EUROPA'S *icy surface shows signs of heat below.*

▲ WHAT LIES BENEATH?
Although the surface is a freezing –274°F (–170°C), heat generated deep in the moon by Jupiter's gravity could have created a hidden ocean where life might flourish.

Methane marvel
In 1997, scientists discovered a new species of centipedelike worm. It was found living on and within piles of methane ice on the seabed of the Gulf of Mexico. If the animal could survive in methane on Earth, could others survive in methane in space?

What an atmosphere
Saturn's largest moon, Titan, has a dense atmosphere—thought to be like the one on early Earth when life began. Titan has the right chemical ingredients for life, including water in the form of ice and carbon compounds that form lakes on the surface. Titan's surface temperature is far too cold for life to survive there, but alien life forms might exist deep underground in hidden lakes of liquid water or ammonia.

▶ WATER OF LIFE? *This false-color radar map shows lakes of liquid methane (a carbon compound) on Titan.*

TERRAFORMING PLANETS

Some NASA scientists think it may be possible to transform lifeless planets into Earth-like planets suitable for humans. This is called terraforming—"forming an Earth." Mars could be terraformed if it were heated up ....

▶ BEFORE
Enough warmth would melt the frozen water and carbon dioxide on Mars, forming oceans and lakes.

▶ AFTER *With enough water, microorganisms and plant life could be brought from Earth to release oxygen into the air and make it breathable.*

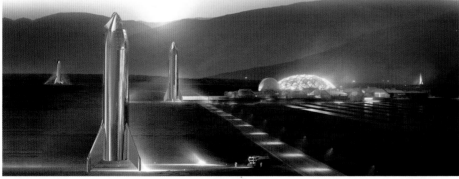

Space colonies
Keeping humans alive for long periods on other worlds will take huge resources, so future astronauts will have to learn to "live off the land." Although the Moon is temptingly close, Mars offers a much better prospect for settlement, thanks to huge reserves of water ice locked in its soil.

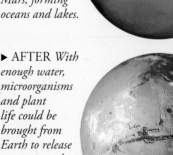

▲ MARTIAN BASE
Ice melted from the Martian permafrost using solar heating could provide drinking water, oxygen, and even rocket fuel for a future interplanetary colony.

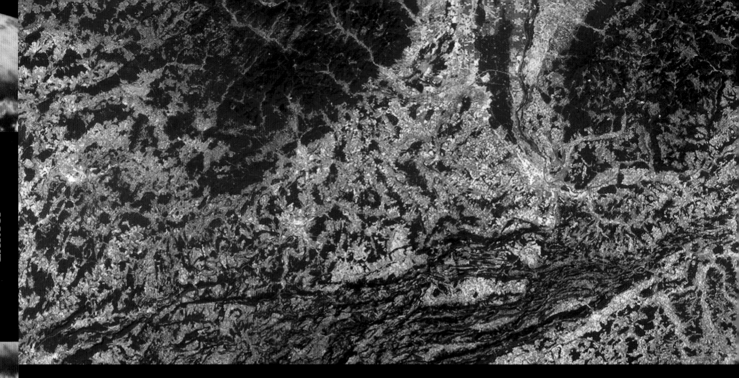

EARTH

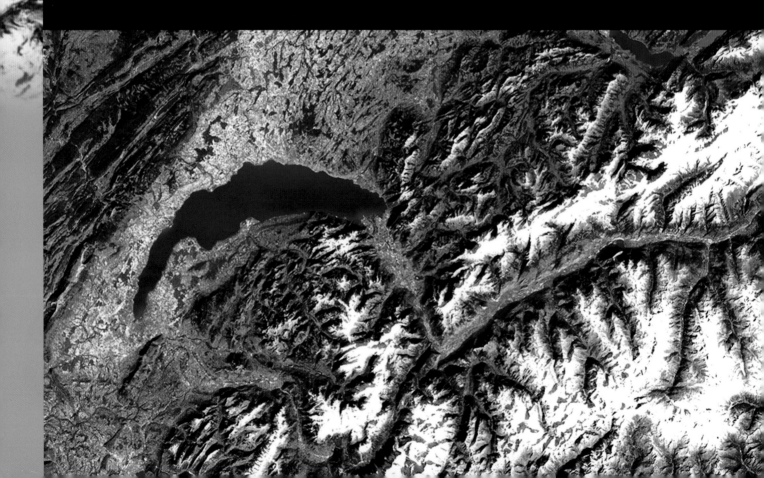

Our home planet is unique. "The third rock from the Sun" is the only world known to have the right conditions for life to flourish—and what an amazing planet it has turned out to be.

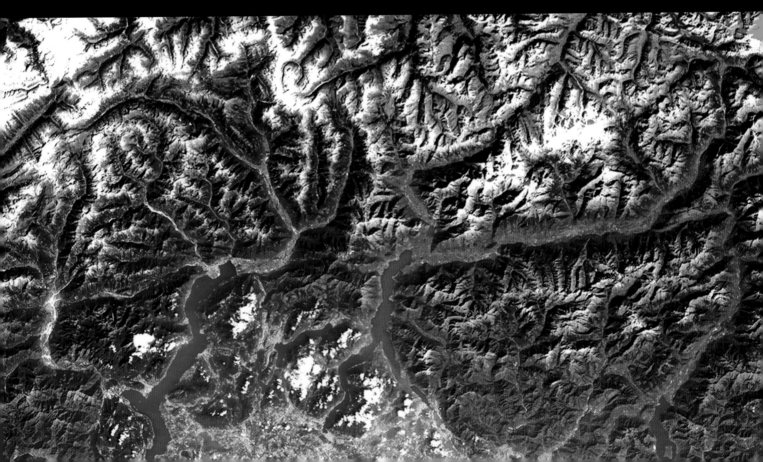

The unique Earth

Earth is a unique planet—the only world known to support any life. It has liquid water on its surface and lots of oxygen. The thick atmosphere protects the surface from radiation and meteorites and the strong magnetic field shields us from harmful particles streaming out from the Sun.

Earth's rocky crust is only about 4 miles (6.5 km) thick under the oceans and about 22 miles (35 km) thick on land.

PLANET PROFILE

- **Average distance from the Sun** 93 million miles (150 million km)
- **Average surface temperature** 59°F (15°C)
- **Diameter** 7,930 miles (12,760 km)
- **Length of day** 24 hours
- **Length of year** 365.26 days
- **Number of moons** 1
- **Gravity at the surface** 1

Crust

Mantle

Inner core

Outer core

The atmosphere is a blanket of gas that surrounds Earth. It is mainly made up of nitrogen (78 percent), oxygen (21 percent), and argon (1 percent).

INSIDE EARTH

Earth has the highest density of any planet in the solar system because its core is mainly made of iron. The very high pressures at the center mean that the inner core remains solid, even at 10,832°F (6,000°C). The outer core is made of molten metal, and the surrounding mantle is a thick layer of partly molten rock. Floating on top of this is a thin, rocky skin called the crust.

Antarctica contains 90 percent of the world's ice and 70 percent of its fresh water. If all of Antarctica's ice melted away, sea levels would rise by more than 200 ft (60 m).

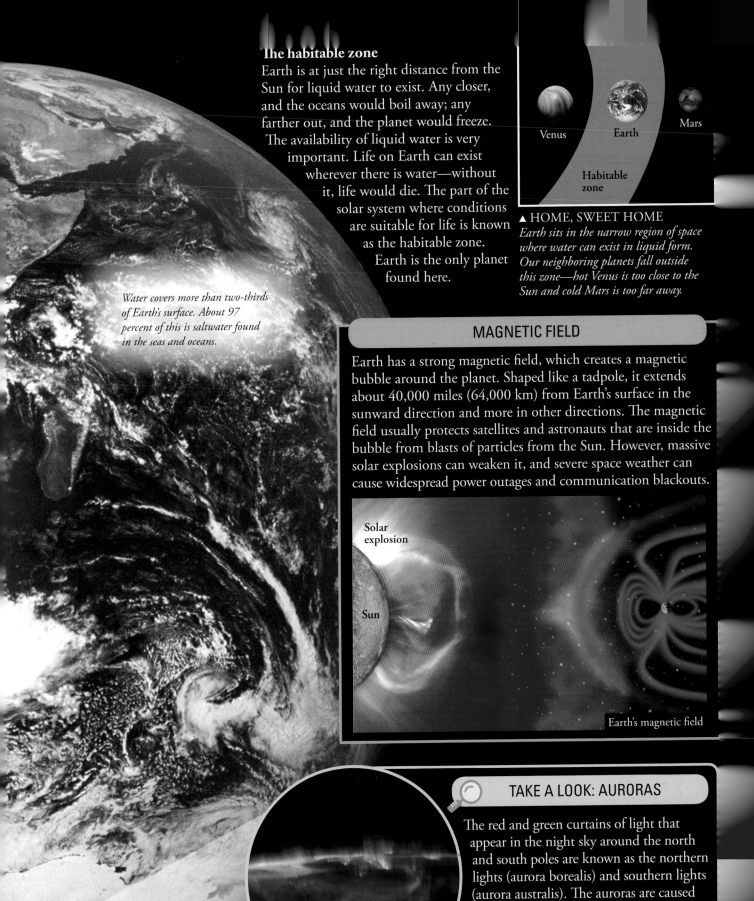

The habitable zone

Earth is at just the right distance from the Sun for liquid water to exist. Any closer, and the oceans would boil away; any farther out, and the planet would freeze. The availability of liquid water is very important. Life on Earth can exist wherever there is water—without it, life would die. The part of the solar system where conditions are suitable for life is known as the habitable zone. Earth is the only planet found here.

Water covers more than two-thirds of Earth's surface. About 97 percent of this is saltwater found in the seas and oceans.

Venus   Earth   Mars

Habitable zone

▲ HOME, SWEET HOME
Earth sits in the narrow region of space where water can exist in liquid form. Our neighboring planets fall outside this zone—hot Venus is too close to the Sun and cold Mars is too far away.

MAGNETIC FIELD

Earth has a strong magnetic field, which creates a magnetic bubble around the planet. Shaped like a tadpole, it extends about 40,000 miles (64,000 km) from Earth's surface in the sunward direction and more in other directions. The magnetic field usually protects satellites and astronauts that are inside the bubble from blasts of particles from the Sun. However, massive solar explosions can weaken it, and severe space weather can cause widespread power outages and communication blackouts.

Solar explosion

Sun

Earth's magnetic field

TAKE A LOOK: AURORAS

The red and green curtains of light that appear in the night sky around the north and south poles are known as the northern lights (aurora borealis) and southern lights (aurora australis). The auroras are caused when high-energy particles from the Sun pour through weak spots in Earth's magnetic field, colliding with atoms in the upper atmosphere and giving off light.

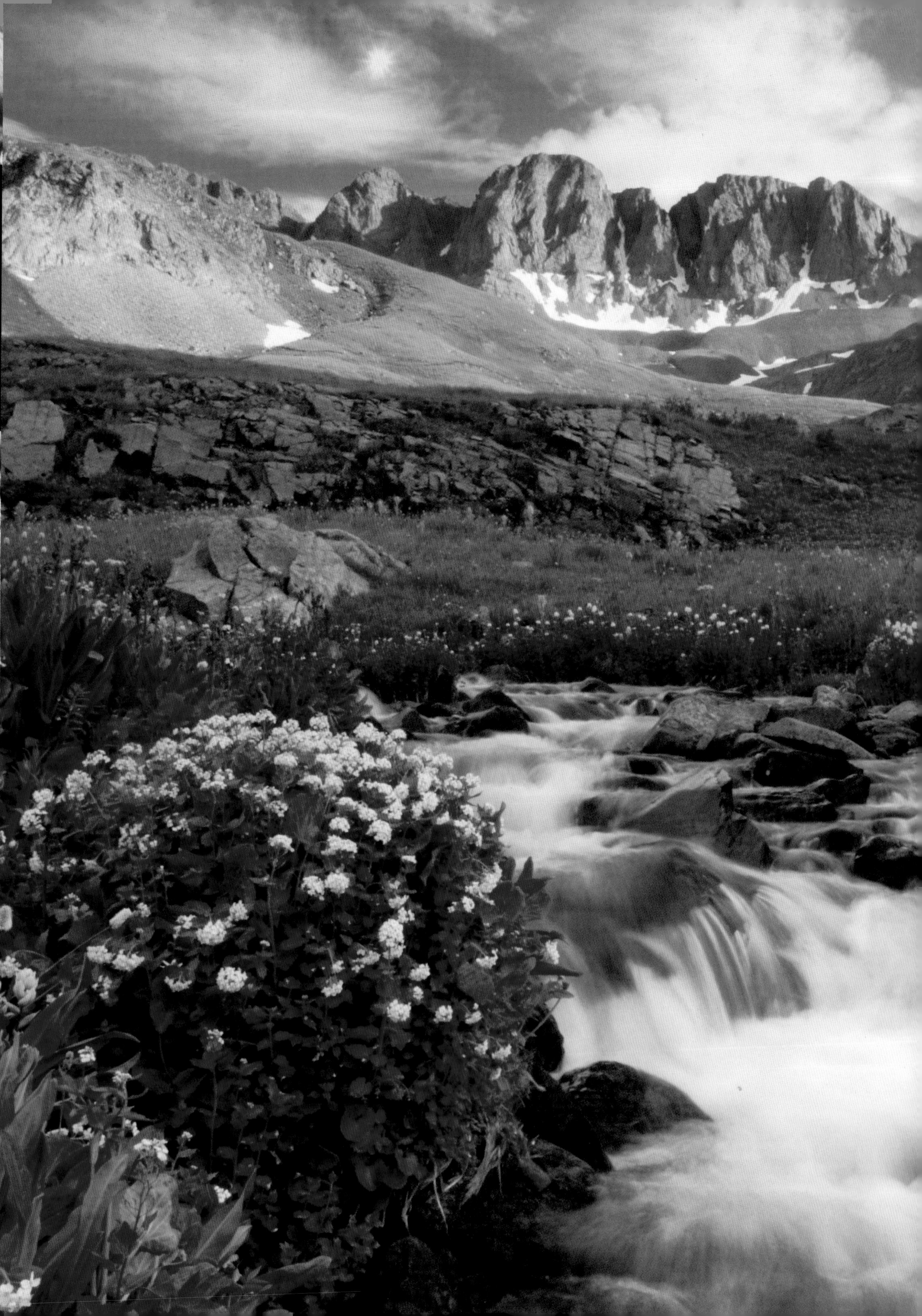

THE PERFECT PLANET

We live on the most amazing rock in the universe. Despite all our efforts to find new, habitable worlds, ours is the only planet so far that has the right conditions for life. Situated at just the right distance from our Sun, it is not too hot nor too cold. The key to life is liquid water, which Earth has in abundance. It drives our weather and makes plants grow, forming the basis of the food chain for animals. Earth is also the only planet we know of that has enough oxygen to keep us alive.

Earth's seasons

We live our lives according to Earth's timetable. Usually, we are awake during the day and asleep at night. The Sun shining on Earth produces day and night. It also plays a role in creating the seasons—spring, summer, fall, and winter.

▲ AN ALIEN'S VIEW *Earth and the Moon appear here in first-quarter phase—half in daylight, half in night.*

EARTH AND MOON

An alien flying past would see the Earth and Moon appearing to change shape. Sometimes the alien would see Earth fully lit, as a bright blue and green disk, sometimes half-illuminated, and sometimes fully in shadow—with various stages in between. The different shapes are called phases. We can see the Moon's phases from Earth.

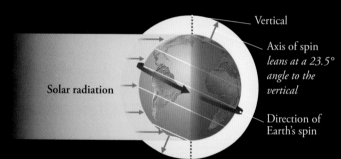

Vertical

Axis of spin *leans at a 23.5° angle to the vertical*

Solar radiation

Direction of Earth's spin

▲ SUNLIGHT INTENSITY *The amount of sunlight received by Earth is affected by the tilt of the axis toward or away from the Sun.*

Day and night

Because Earth is tilted as it spins, the period of daylight changes throughout the year, unless you live on the equator. The polar regions experience this to the extreme, with very long days in summer and very long nights in winter. North of the Arctic Circle and south of the Antarctic Circle, the Sun does not rise in midwinter or set in midsummer. Because of this, areas such as northern Norway and Alaska are known as the "land of the midnight Sun."

▲ MIDNIGHT SUN *This multiple-exposure image shows how the Sun dips toward the horizon but never sets below it in a polar summer.*

ON A TILT

The seasons are caused by Earth rotating at a slight angle, like a spinning top that has been knocked slightly to one side. If Earth were to spin upright, we would not have any seasons.

Most planets rotate at a tilt, but if they lean too much, the seasons can be very strange. Summers and winters on Uranus each last for 21 years.

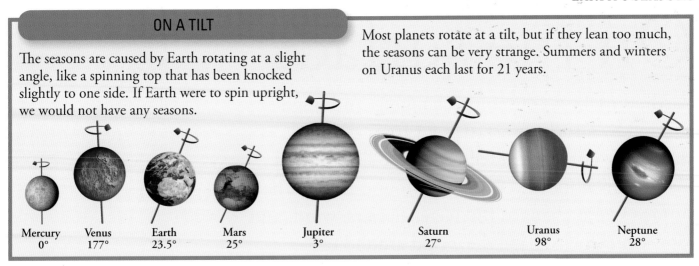

| Mercury 0° | Venus 177° | Earth 23.5° | Mars 25° | Jupiter 3° | Saturn 27° | Uranus 98° | Neptune 28° |

SEASONS

Those not living near the poles or the equator experience four seasons—spring, summer, fall, and winter. At the equator, the period of daylight hardly changes and the Sun is high in the sky, so it is always warm. Our spinning Earth is tilted at 23.5 degrees to the plane of its orbit. When the north pole is tilted toward the Sun, it is summer in the northern hemisphere and winter in the southern hemisphere. When the north pole is tilted away from the Sun, it is winter in the northern hemisphere and summer in the southern hemisphere.

Sun

Earth

Southern summer occurs when the north pole tilts away from the Sun.

Day   Night

Northern summer occurs when the north pole tilts toward the Sun.

▲ EARTH'S ORBIT *Earth moves around the Sun in an oval-shaped orbit, which varies Earth's distance from the Sun but is not responsible for the seasons.*

▼ VEGETATION *patterns (green) change according to how much light is received in each season.*

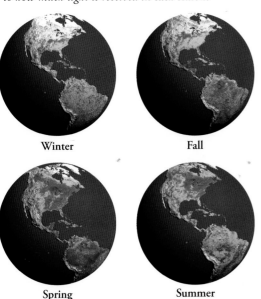

Winter   Fall

Spring   Summer

The seasonal Sun

Earth's temperature is influenced by the length of the day and by the seasons. In summer, the Sun is above the horizon for longer and higher in the sky. Less heat is absorbed by the atmosphere and more heat is absorbed by the ground and oceans. In winter, the Sun is above the horizon for a shorter length of time. During the long nights, more heat escapes to space than is provided by the Sun during the day.

► IN HOT WATER
This map shows how sunlight affects sea temperatures around the world, with warm waters in red around the equator, cooling through orange, yellow, and green. Cold waters are shown in blue.

171

On the surface

Earth's surface is constantly changing. Although covered by a rocky crust, it is far from stiff and static. The crust is divided into huge slabs, called plates, which move very slowly around Earth. The surface is also changed by rivers, glaciers, wind, and rain, which help shape the world around us.

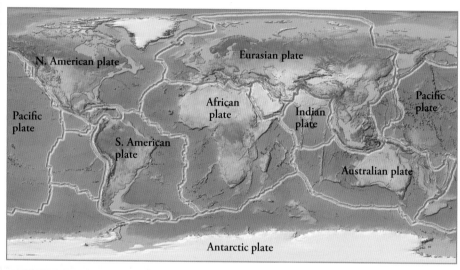

N. American plate

Eurasian plate

African plate

Indian plate

Pacific plate

Pacific plate

S. American plate

Australian plate

Antarctic plate

EARTH'S PLATES

The rocky plates that make up the crust float on Earth's dense mantle. They move between 1 and 6 inches (3 and 15 cm) a year, changing the positions of the continents over time. Some plates move apart, others slide toward or past each other. Their movements build mountain ranges and cause earthquakes, tidal waves, and volcanic eruptions.

Earthquakes and volcanoes

The edges of plates are dangerous places to live. Major earthquakes occur where plates collide and cities such as San Francisco or Tokyo, which lie near active plate boundaries, suffer from frequent, large earthquakes. Many volcanoes occur at plate boundaries, where one plate slides under another, allowing molten rock to escape to the surface.

Mountain ranges

Most continents have mountain ranges. These occur where two plates collide, pushing the crust up to form high peaks. Standing at 29,029 ft (8,848 m), Mount Everest is the highest mountain in the world. It is part of the Himalayan mountain range that formed when the Indian plate crashed into the Eurasian plate. There are also volcanic mountains that rise from the seabed. The tallest of these is Mauna Kea, an inactive volcano in Hawaii. Measured from the ocean floor, Mauna Kea is even taller than Everest.

WATER

Water world

As streams and rivers flow down from high ground, they pick up sediment and small rock fragments. These abrasive particles grind away at the landscape. Over time, this process wears away mountainsides and carves out deep canyons. Rivers can also build up and create new landscape features by depositing mud and silt as they approach the sea. The sea itself is a massive force of change—the waves grind away at cliffs and shorelines, changing coastlines and forming spectacular shapes in the rocks.

▼ WILD, ROCKY LANDSCAPE *of wind-eroded red sandstone in Colorado.*

Windswept In dry places with little water or plant life, wind is the major source of erosion. The wind blasts rocks at high speed, carrying away loose particles of rock and grinding these against existing landscape features. Over years, this wears down rocks and can produce some amazing shapes—arches, towers, and strange, wind-blown sculptures.

WIND

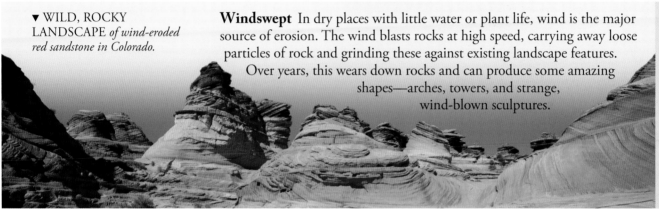

ICE

Rivers of ice

Glaciers are large, moving sheets of ice that occur at the poles and high in mountain ranges. Some barely move, while others surge forward, traveling as fast as 65–100 ft (20–30 m) a day. These rivers of ice dramatically alter the landscape, eroding rock, sculpting mountains, and carving out deep glacial valleys. Glaciers pick up rocks and debris, dragging them along and leaving holes or depressions in the valley floor. As the glaciers melt, they produce lakes and leave boulders strewn across the landscape.

◄ SAN ANDREAS FAULT
San Andreas in California is a fault, or crack, in the Earth's crust where two plates, the Pacific and the North American, are sliding past each other. On average, they move only an inch or so each year. This motion is not consistent—the plates remain locked together until enough stress builds up and a slip occurs. The sudden movement of the plates releases energy and causes earthquakes.

Up in the air

Life could not survive on Earth without the thick blanket of gases known as the atmosphere. The atmosphere protects us from harmful radiation and small incoming meteorites. It also provides us with our weather and helps keep Earth warm.

OZONE HOLE

The atmosphere contains a form of oxygen known as ozone. Ozone is important because it helps block harmful ultraviolet radiation coming from the Sun. In 1985, a hole in the ozone layer was found over Antarctica and a smaller hole was found over the Arctic a few years later. These holes were caused by the release of human-made chemicals called chlorofluorocarbons (CFCs). These chemicals are now banned, but the ozone holes are likely to remain for many years and are closely watched by satellites in space.

ATMOSPHERIC ZONES

EXOSPHERE

Satellite

375 miles (600 km)

THERMOSPHERE

Northern lights

50 miles (80 km)

MESOSPHERE

Shooting stars

30 miles (50 km)

STRATOSPHERE

Airplanes

5–10 miles (8–16 km)

TROPOSPHERE

Clouds

Other gases

Oxygen

Nitrogen

The sky appears blue because blue light is scattered more than other colors by the gases in the atmosphere.

IT'S ALL A GAS

The atmosphere extends about 600 miles (1,000 km) into space. It is thickest near the ground and quickly becomes thinner as you move upward. The most common gases in the atmosphere are nitrogen (78 percent) and oxygen (21 percent). Other gases include argon, carbon dioxide, and water vapor.

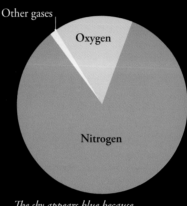

◄ ZONES *Earth's atmosphere consists of five layers. The layer closest to the ground is the troposphere. All our weather occurs in this layer. The stratosphere is more stable and also contains the ozone layer. Although the air is much thinner in the mesosphere, there is enough to cause meteors to burn up on entry. Auroras occur in the thermosphere. The exosphere marks the upper limit of the atmosphere, where most spacecraft orbit.*

WATER CYCLE

The water cycle is a continuous movement of water between Earth's surface and its atmosphere. It is powered by heat from the Sun and provides us with a constant source of fresh water.

High above the ground, the water vapor cools. It turns back into droplets of water and clouds are formed.

When the droplets get too heavy, they fall back to the surface as rain or snow.

Water in the rivers and oceans is heated by the Sun and evaporates, turning into a gas called water vapor.

Some water soaks into the ground to form groundwater.

The rest of the water runs off the land, flowing into streams and rivers.

About 90 percent of the evaporated water that enters the water cycle comes from the oceans.

Streams and rivers channel water back into lakes or toward the ocean.

TAKE A LOOK: CLOUDS AND WEATHER

Stratus clouds

Cumulus clouds

Thunder clouds

Earth's weather takes place in the troposphere, where water vapor cools to form clouds. There are many types of clouds. Stratus clouds form wide layers in still air. Cumulus clouds bubble up where warm air rises. Rapidly rising air carries clouds to great heights and large, tall clouds called cumulonimbus often produce rain and sometimes hailstones. Cirrus clouds at the very top of the troposphere are made of tiny crystals of ice.

◄ SUPER STORM *The rarest type of thunderstorm is the supercell. It produces the most violent weather, including deadly lightning, giant hail, flash floods, and tornadoes.*

Storm forces

Hurricanes are the most powerful storms on Earth. Storms over tropical waters become hurricanes when wind speeds reach more than 74 mph (120 kph). Hurricanes in the southern hemisphere spin in a clockwise direction, while those in the northern hemisphere spin counterclockwise.

▲ EYE OF THE STORM *The air at the center of a hurricane (the eye) remains still while powerful winds rage around it.*

▼ DUST STORMS *are caused by strong winds passing across deserts or dry, dusty areas. They can pick up thousands of tons of sand or dust. An approaching storm can appear as a solid wall, reaching up to 1 mile (1.6 km) from the ground.*

Life on Earth

Earth is the only place we know where life exists. Life is found almost everywhere on the planet—from the highest mountains to the deepest ocean trenches. It is even found in boiling hot springs and inside solid rock.

THE ORIGINS OF LIFE

The first simple life-forms probably appeared on Earth about 3.8 billion years ago. No one knows how life began but scientists think it may have started in the oceans, since the land was very hot and the atmosphere was poisonous. Others think comets or meteors brought complex chemicals from outer space. However it began, simple molecules formed and began to copy themselves, eventually growing into cells, and then colonies. Over time, these evolved into more complicated organisms that began to colonize the land.

Life begins

The first life-forms were simple, single cells that probably lived in the oceans and hot springs. Over billions of years, single-celled organisms became a lot more complex and multicellular life evolved.

Early cell

TIMELINE OF LIFE ON EARTH

EARLY EARTH

4.5 billion years: Earth forms

3.6 billion years: Blue-green algae release oxygen into atmosphere

3.8 billion years: Simple bacteria appear in the oceans

2.2 billion years: First complex organisms, the ancestors of animals, plants, and fungi, appear

FIRST LIFE

635 million years: First complex animals appear in the sea

530 million years: Fish evolve

470 million years: First plants colonize land

370 million years: Seed plants appear

395 million years: Tetrapods take first steps onto dry land

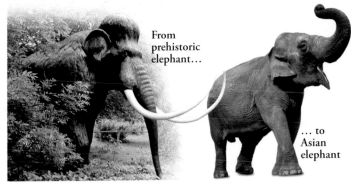

From prehistoric elephant…

… to Asian elephant

Evolution

Earth supports many forms of life, including plants, animals, and tiny bacteria. All living things have adapted to their surroundings through a process called evolution. This takes place through processes that change the genes of living organisms over many generations.

Extinctions

At various times during Earth's history, many life-forms have been wiped out. Some mass extinctions were probably caused by huge volcanic eruptions belching out clouds of gas and ash. These would have blocked out the Sun, causing the temperature to drop and killing many of the plants that animals needed for food. The extinction of the dinosaurs 66 million years ago has been blamed on volcanic eruptions triggered by an asteroid impact.

▲ TIKTAALIK
This extinct lobe-finned fish lived during the Late Devonian period, 375 million years ago.

EARTH

Hydrothermal mussels and shrimp

Black smoker

Giant tube worm

Black smokers

Most plants and animals rely on sunlight to survive, but some deep-sea creatures live in total darkness. Thousands of feet below the surface, water escapes from the super-hot mantle through cracks in the rock. These hot volcanic vents, or "black smokers," are home to dense communities of giant tube worms, mussels, shrimp, and crabs. They live on bacteria that are able to harness energy from chemicals dissolved in the hot water. Some bacteria also live inside solid rock or on cold parts of the ocean floor and get their energy by eating the minerals in the rock.

TAKE A LOOK: OCEAN BLOOMS

The oceans are not just home to large creatures, such as fish and whales. Among the most important forms of ocean life are microscopic plants called phytoplankton. These tiny organisms float in the surface waters where there is plenty of sunlight. They provide an important source of food for a range of animals, from small shrimp to huge whales. When a great number of phytoplankton are concentrated in one area, they change the color of the ocean's surface. Sometimes these "blooms" are so big they can be seen from space.

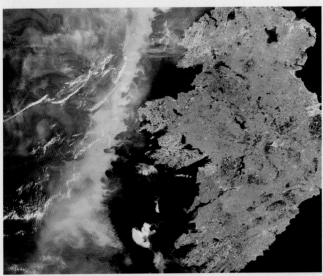

▲ PLENTIFUL PLANKTON *A turquoise-colored phytoplankton bloom appeared off the coast of Ireland in June 2006.*

MORE COMPLEX LIFE-FORMS

314 million years: Winged insects take to the skies

225 million years: First mammals emerge

231 million years: Dinosaurs and birds evolve from reptiles

220 million years: Reptiles begin to fly (pterosaurs)

66 million years: Mass extinction wipes out dinosaurs and many other life-forms

MODERN TIMES

60 million years: Mammals take over the world and modern forms of fish, reptiles, plants, and insects appear

300,000 years: Modern man (*Homo sapiens*) evolves

7 million years: Apes descend from the trees and start walking upright

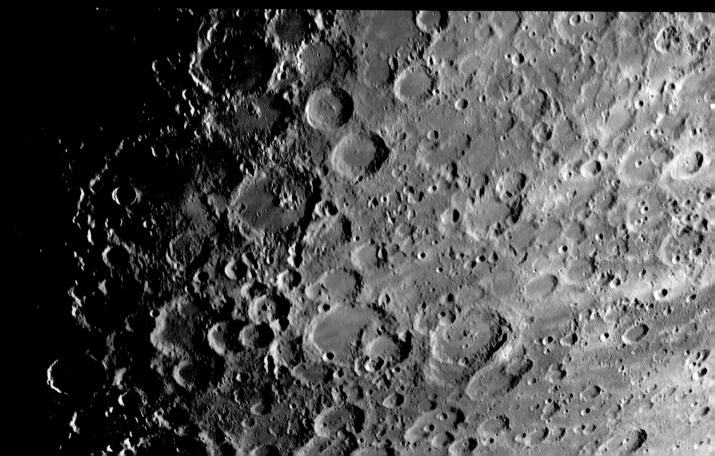

THE
MOON

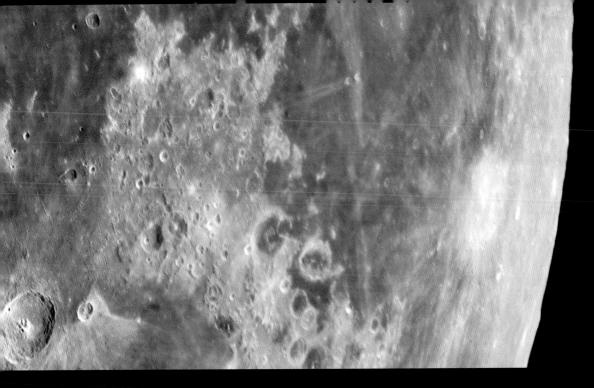

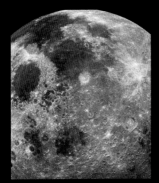

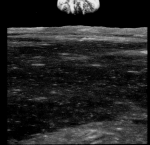

A full Moon is the second-brightest object in the sky, after the Sun. Our Moon was humankind's first destination in space, but only 12 people have ever walked on its surface.

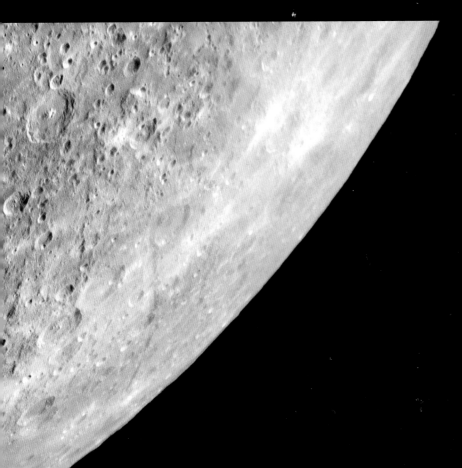

Earth's companion

Earth and the Moon have been close partners for about 4.5 billion years. Although the Moon is much smaller than Earth, it influences our planet in many ways and has fascinated humans for thousands of years.

Lunar tides

Tides are created by the Moon's gravity pulling on Earth's water. At any one time, there is a place on Earth that is nearest to the Moon and one that is farthest away. Here the seas "bulge" out, creating high tides. The bulges move around the Earth as it rotates.

▲ LOW TIDES *occur twice a day when a place is at right angles to the Moon's gravity pull.*

▲ HIGH TIDES *happen twice daily, too, when a place is aligned with the Moon.*

Solar tides

The Sun also has a weak effect on tides. When the Moon, Earth, and Sun align, their combined gravity causes very low and very high spring tides. When the Moon and Sun are at right angles, you have a neap tide.

▲ SPRING TIDES *occur when solar and lunar tides join forces to create an extremely strong gravitational pull.*

▲ DURING A NEAP *tide the high tide is slightly lower than usual and the low tide is slightly higher than usual.*

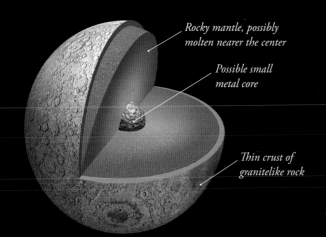

Rocky mantle, possibly molten nearer the center

Possible small metal core

Thin crust of granitelike rock

MOON PROFILE

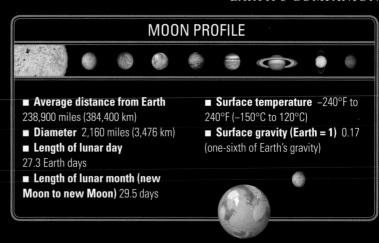

- **Average distance from Earth** 238,900 miles (384,400 km)
- **Diameter** 2,160 miles (3,476 km)
- **Length of lunar day** 27.3 Earth days
- **Length of lunar month (new Moon to new Moon)** 29.5 days
- **Surface temperature** –240°F to 240°F (–150°C to 120°C)
- **Surface gravity (Earth = 1)** 0.17 (one-sixth of Earth's gravity)

INSIDE THE MOON

The Moon has a crust of brittle rock about 30 miles (50 km) thick that is riddled with cracks. Beneath the crust is a deep mantle that is thought to be rich in minerals, similar to those found in Earth rock. The mantle may extend all the way to the center, or the Moon may have a small metal core.

Slowing down
Tidal forces between the Earth and Moon are gradually slowing down Earth's rotation, making the day longer. When Earth was formed, a day lasted only six hours. By 620 million years ago, a day had lengthened to 22 hours. Eventually, tidal forces will increase our day length to 27.3 Earth days, matching the lunar day exactly.

In a spin
The Moon takes 27.3 days to orbit Earth once, but also 27.3 days to spin once on its axis. As a result, it keeps the same side facing Earth—the "near side." Even so, variations in the Moon's orbit allow parts of its far side to come into view now and then. Tidal forces between the Earth and Moon are causing the Moon to move slowly away from Earth by 1½ in (3.8 cm) a year.

New Moon

Crescent Moon waning

Crescent Moon waxing

AS THE MOON moves from new Moon to full Moon it is said to be "waxing." As it moves from a full Moon through to the next new Moon it is said to be "waning." When more than half of the Moon's face is visible it is described as "gibbous."

Last quarter

First quarter

Gibbous Moon waning

Gibbous Moon waxing

Full Moon

▼ VIEW OF the Earth and the Moon looking down onto their north poles.

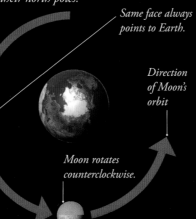

Same face always points to Earth.

Direction of Moon's orbit

Moon rotates counterclockwise.

PHASES OF THE MOON

For centuries, people have been fascinated by the way the Moon goes through a cycle of "phases" that repeats every 29.5 days. These phases occur because we see different amounts of the Moon's sunlit side as the Moon orbits Earth.

Eclipses

Eclipses are among the most spectacular astronomical events you can see. They occur when the Earth, Moon, and Sun all line up so that Earth casts a shadow on the Moon or the Moon casts a shadow on Earth. The Sun or Moon appear to go dark to people standing inside these shadows.

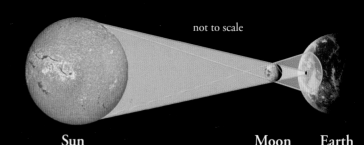

not to scale

Sun   **Moon**   **Earth**

▲ MOON BLOCK *A total solar eclipse occurs when the Moon completely blocks the light from the Sun. All that can be seen is the corona (the Sun's atmosphere) as a shimmering halo of light around it.*

▶ MASKED BY THE MOON *As the Moon passes in front of the Sun, we see less and less of the Sun's disk.*

▶ DIAMOND RING
At the start and end of a total eclipse, sunlight shining through lunar mountains can create the stunning "diamond ring" effect.

Shadow play
A total solar eclipse can be seen only from the center of the Moon's shadow—the umbra. The umbra sweeps across Earth during an eclipse, tracing a path thousands of miles long but only about 60 miles (100 km) wide. Outside the umbra, the Moon casts a partial shadow, causing a partial solar eclipse.

SOLAR ECLIPSES

The Moon passes between the Sun and Earth every month at "new moon," but because its orbit is slightly tilted, it usually does not pass directly in front of the Sun. Occasionally, however, it does move directly in front of the Sun and causes a solar eclipse. Although the Sun is 400 times wider than the Moon, by a curious coincidence, it is also 400 times farther away. As a result, when viewed from Earth, the Moon's disk fits exactly over the Sun's disk during a total solar eclipse.

WATCH THIS SPACE

… Carefully! When viewing a solar eclipse, you shouldn't look directly at the Sun without proper eye protection. Only when the Sun is fully eclipsed and the faint corona is visible, is it safe to look directly at it.

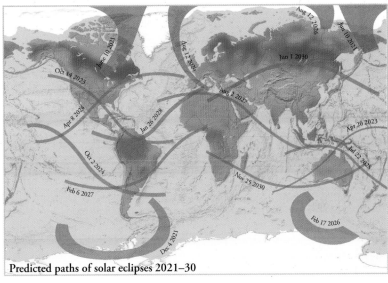

Predicted paths of solar eclipses 2021–30

When day becomes night

A total solar eclipse occurs about every 18 months. If you are in the right place to see one, it is an amazing experience. As the last rays of sunlight are eclipsed, darkness falls, stars appear, and day turns to twilight. All that can be seen of the Sun is its hazy outer atmosphere.

LUNAR ECLIPSES

Up to three times a year, the Moon passes through Earth's enormous shadow and a lunar eclipse occurs. The Moon does not become completely black as some sunlight is refracted (bent) by Earth's atmosphere, which makes the Moon turn orange-red, like a red sunset. In the 18th century, Chinese astronomer Wang Zhenyi studied the relationships between the Earth, the Sun, and the Moon and accurately explained the causes of a lunar eclipse.

▼ A LUNAR ECLIPSE *When Earth comes between the Sun and the Moon, the Moon is in shadow.*

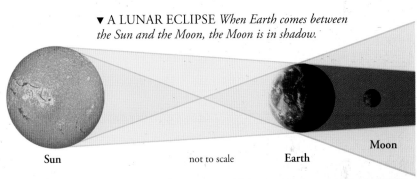

Sun   not to scale   **Earth**   **Moon**

WHEN AND WHERE TO SEE A LUNAR ECLIPSE	
May 26, 2021	*east Asia, Australia, Pacific, Americas*
November 19, 2021	*Americas, north Europe, east Asia, Australia, Pacific*
May 16, 2022	*Americas, Europe, Africa*
November 8, 2022	*Asia, Australia, Pacific, Americas*
May 5, 2023	*Africa, Asia, Australia*
October 28, 2023	*east Americas, Europe, Africa, Asia, Australia*
March 25, 2024	*Americas*
September 18, 2024	*Americas, Europe, Africa*
March 14, 2025	*Pacific, Americas, west Europe, west Africa*
September 7, 2025	*Europe, Africa, Asia, Australia*
March 3, 2026	*east Asia, Australia, Pacific, Americas*

▲ RED MOON *This time-delay photograph shows the stages of a single lunar eclipse. Earth's shadow can take four hours to move across the Moon, but "totality"—when the Moon is fully inside the shadow—lasts only around one hour.*

The *lunar surface*

Even with the naked eye, we can see surface features on the Moon. The dark areas are called "maria," the Latin word for seas, because early astronomers mistook them for oceans. The Italian scientist Galileo was the first person to view the Moon with a telescope and was amazed to see mountains, plains, and valleys.

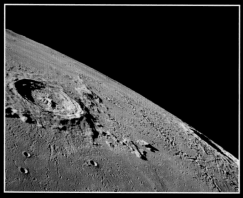

▲ MOON CRATERS *vary in size from a few inches to about 186 miles (300 km) in diameter. The larger craters often have central mountains where the crust rebounded after impact, as in the 36 mile (58 km) wide crater Eratosthenes. It is surrounded by rays of material thrown out from the nearby crater Copernicus.*

Maria

Thousands of craters pepper the Moon's face like scars, evidence of violent clashes with asteroids and comets.

Highlands

LUNAR HIGHLANDS

The cratered areas outside the maria are called highlands. These cover most of the Moon's surface, especially on the far side. The highland rock is chemically different from the maria rock and lighter in color. The lunar mountains that line the edges of craters or maria reach more than 2 miles (3.5 km) in height and are smoother than Earth mountains. The surface is covered in rocks and powdered gray dust several feet deep.

THE FAR SIDE

We only ever see one side of the Moon from Earth, so our first view of the far side came from pictures taken by the Soviet probe Luna 3 in 1959. Later, NASA Apollo missions took even clearer pictures, such as the one shown here, centered on the boundary between the near and far sides. The far side has few maria and consists mostly of heavily cratered highlands.

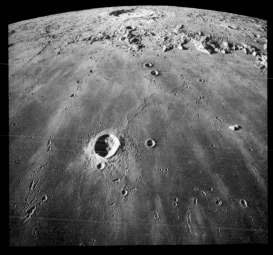

Waterless seas

The lunar maria, or "seas," are flat plains of volcanic rock. Astronomers think they formed during the Moon's first 800 million years, when molten rock welled up and filled the bottoms of gigantic basins. The lava cooled and solidified to form smooth plains. After the maria formed, the rate of meteorite impacts dropped, so the maria have fewer craters than the much older highlands.

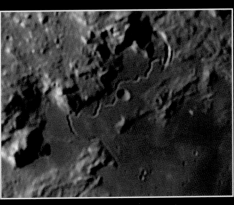

▲ LAVA FLOWS *Snakelike channels were formed by rivers of lava billions of years ago. The top of the cooling lava formed a solid roof. Later, the liquid lava drained away and the roof of the tunnel collapsed, leaving winding channels known as rilles.*

📷 WATCH THIS SPACE

This dusty footprint will remain on the Moon forever, as there is no wind to blow it away. Moon dust is said to smell like gunpowder. The fine dust particles covered the astronauts' spacesuits and equipment when they stepped outside.

Destination Moon

Humanity's dream of space travel became a reality in the 1950s and 1960s, when the Russians and Americans set out to be the first to conquer space. In the end, both countries scored space firsts: the USSR with uncrewed probes and the first person in space, and the US with a person on the Moon.

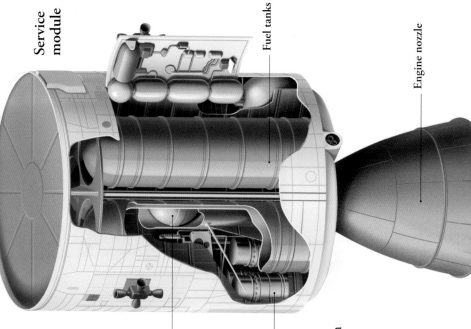

Command module

Astronauts' seats

Forward heat shield

Instrument panel

Service module

Fuel tanks

Engine nozzle

Helium tanks

Fuel cells

WATCH THIS SPACE

By the 1950s, a lunar mission had become a real possibility due to advances in space technology. Many toys, books, and films from this time are based on space travel.

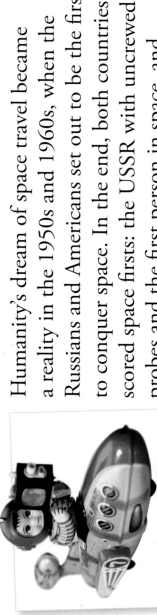

To the Moon and back

The first mission to land people on the Moon began from Cape Canaveral in Florida on July 16, 1969, when a Saturn V rocket sent the Apollo 11 spacecraft on its historic journey. Actually, the dream almost didn't happen—the lunar module touched down on the Moon with less than 30 seconds of fuel remaining as its pilot, Neil Armstrong, struggled to find a safe landing site.

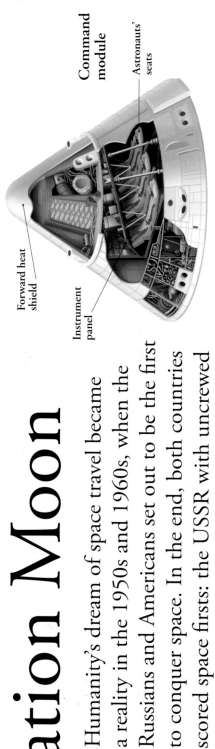

4. LM lands on lunar surface.

5. Upper stage of LM returns to lunar orbit to dock with CSM.

3. LM separates for landing. CSM stays in lunar orbit with fuel for return to Earth.

2. Rocket is discarded. CSM and LM proceed into orbit around the Moon.

6. CSM fires its rockets to return to Earth.

7. Command module separates from service module and returns crew to Earth.

1. Command and service module (CSM) and lunar module (LM) are sent into Earth orbit.

READY, STEADY, GO!

Over 100 spacecraft have been sent to the Moon since the first lunar mission in 1959, although many of them were failures. Here are some of the early highlights.

Lunar module upper stage

Docking tunnel

Equipment bay

Oxygen tank

Fuel tank

Rendezvous radar antenna

Control console

Lunar module landing stage

Fuel tank

Lunar surface sensing probe

Scientific experiments package

Landing pad

Exit platform

Apollo 11 exploded

The spacecraft consisted of three modules: the command module (CM) for the astronauts to live, work, and ultimately return to Earth in; the service module (SM) containing fuel and equipment for supplying the astronauts with water, electricity, and oxygen; and the two-stage lunar module (LM) for the all-important Moon landing.

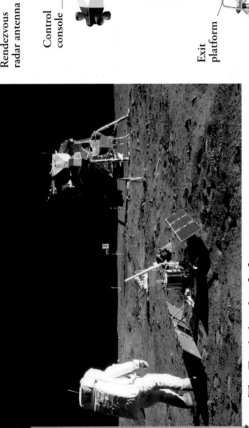

The Eagle has landed

The lunar module was nicknamed the Eagle. Under its thin aluminum exterior were gold-coated thermal blankets to protect it against the huge temperature changes. Once they had landed safely, the astronauts donned their extravehicular activity spacesuits and went out onto the surface to carry out some scientific experiments.

▲ THIS VIEW *from the Apollo 11 spacecraft shows Earthrise over the Moon's horizon. The lunar landscape is the area of Smyth's Sea on the near side of the Moon.*

January 1959
Soviet probe Luna 1, the first spacecraft sent to the Moon, malfunctions and misses the Moon by 3,700 miles (6,000 km).

September 1959
Luna 2 makes a deliberate crash-landing, becoming the first craft to touch down on the Moon.

October 1959
Luna 3 becomes the first craft to photograph the far side of the Moon.

July 1964
US probe Ranger 7 takes thousands of photos of the Moon's surface before deliberately crash-landing.

February 1966
Luna 9 becomes the first craft to make a soft landing on the Moon.

April 1967
US probe Surveyor 3 lands on the Moon and photographs the future landing site of the Apollo 12 crewed mission.

December 1968
Humans orbit the Moon for the first time during NASA's Apollo 8 mission.

July 1969
Neil Armstrong and Buzz Aldrin are the first people ever to set foot on the Moon as part of NASA's Apollo 11 mission.

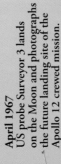

November 1970
Lunokhod 1, a Russian rover looking like an eight-wheeled baby carriage, becomes the first vehicle to drive on the Moon.

Man on the Moon

▲ TRANQUILITY BASE *Aldrin and Armstrong's space walk was televised around the world.*

On July 20, 1969, 500 million people watched on TV as Neil Armstrong became the first person to set foot on the lunar surface, announcing, "That's one small step for [a] man, one giant leap for mankind." Altogether, 12 people walked on the Moon between 1969 and 1972 in six successful missions.

Walking on the Moon

Because the Moon has such low gravity, the astronauts weighed only one-sixth of their normal weight on the Moon—and so did their heavy life-support backpacks, which became much easier to carry. However, the low gravity made it impossible to walk normally. Some used a "kangaroo hop," others a loping walk. Some even enjoyed "skiing" or gliding over the Moon dust by pushing off with their toes.

That's rubbish!

The Moon is littered with lunar modules, flags, probes, and other pieces of equipment that have been left there or that have crash-landed—planned or unplanned! The uncrewed Russian spacecraft Luna 15 crashed into the Moon just hours after Apollo 11's lunar module landed.

Storage for tools, lunar rock, and soil samples

Camera

Dish antenna for relaying pictures back to Earth

Moon buggy

Apollo missions 15 to 17 carried a 10 ft (3 m) long, open-topped roving vehicle that was carried, folded up, on the side of the lunar module. The battery-powered rover had a top speed of 11.5 mph (18.6 kph).

Wire-mesh tires

The Apollo astronauts brought back case upon case of rock and soil samples from their six missions. Despite the low gravity, it was hard, dirty work. The astronaut's arm and hand muscles tired very quickly in the restrictive spacesuits and gloves. Bending over was almost impossible, so the astronauts had special tools to pick up rocks. They also found out that Moon dust was powdery, very abrasive, and extremely clingy, turning spacesuits gray, scratching visors, and even wearing through the surface layers of their boots.

▲ ROCK SAMPLES *are studied to help scientists piece together the Moon's history. This basalt rock was found by Apollo 15 astronauts and shows that the Moon had a volcanic past.*

▲ TRAINING *missions were carried out on Earth to test tools and maneuvers. Here, the astronauts are practicing in a volcanic crater in Arizona.*

📷 WATCH THIS SPACE

As a memento of his trip, Apollo 16 pilot Charles Duke left a photo of his family and a medal in a plastic bag on the lunar surface. The back of the photo is signed by his family.

Let it shine

One of the scientific experiments that the Apollo astronauts set up on the lunar surface was a laser reflector. Scientists back on Earth aimed a laser at the Moon and then measured how long it took for the reflection to come back. From these measurements, they discovered that the Moon is slowly drifting away from Earth by 1½ in (3.8 cm) a year.

◄ SCIENTISTS *sent the laser beam through an optical telescope at the McDonald Observatory in the US. This experiment determined the distance between Earth and the Moon to an accuracy of 1 in (2.5 cm).*

▲ SEVERAL *reflectors have been placed on the Moon since 1969. The returning beams from the laser reflectors are too weak to be seen with the human eye, and sensitive amplifiers are used to enhance the signal.*

⬛ SPLASHDOWN!

After a fiery reentry into the Earth's atmosphere, parachutes helped the Apollo command module's descent into the Pacific Ocean. The water cushioned the landing, and once down, floats were activated to keep the capsule upright.

▲ PARACHUTES *ensured a safe landing for the cone-shaped command module.*

▲ FROGMEN *helped the crew from their charred capsule and into life rafts before airlifting them aboard a navy ship.*

▲ THE APOLLO 11 *crew spent several weeks quarantined in an airtight container when they returned to make sure they hadn't picked up any alien bacteria.*

ALMOST THERE
*This picture of the Apollo 11
command module was taken by the
Eagle lander as it began its descent
to the Moon's surface, leaving pilot
Michael Collins to orbit alone.*

Return to the Moon

After the Apollo program ended in 1972, and the last Luna probe visited the Moon in 1976, there were no missions until Japan's Hiten in 1990. Today, space agencies around the world are planning future missions to the Moon and beyond.

◀ *Japan achieved its first-ever lunar flyby, lunar orbiter, and lunar surface impact with Hiten—only the third nation ever to achieve this.*

MAPPING MISSIONS

The launch of the Clementine spacecraft in 1994 heralded NASA's return to the Moon. Over the course of its 71-day orbit, Clementine mapped all of the 15 million square miles (38 million square km) of the Moon. NASA followed up this successful mission with the Lunar Prospector in 1998 and the LRO in 2009.

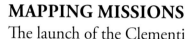

Multitalented

Clementine not only carried equipment into outer space to test how it coped in space, but it also mapped the topography (height) of the Moon's surface and the thickness of its crust, taking over a million pictures in total. Data provided by Clementine suggested that there may be frozen water in the deep craters near the south pole.

▲ CLEMENTINE *bounced radio waves off the Moon's surface and found the first evidence of water ice.*

Lunar Reconnaissance Orbiter (LRO)

The uncrewed LRO was launched in 2009 to investigate possible sites for setting up a crewed base on the Moon. The Lunar Crater Observation and Sensing Satellite (LCROSS) was sent up at the same time. It was crashed into the surface in a search for water ice.

▶ THE LCROSS *mission confirmed that there was a little water ice in at least one of the Moon's craters. LRO's pictures also disproved claims that the Apollo missions were a hoax.*

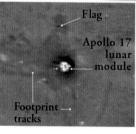

Flag

Apollo 17 lunar module

Footprint tracks

GRAIL

The Gravity Recovery and Interior Laboratory was a pair of NASA satellites launched into lunar orbit in 2011. Flying two satellites in formation allowed scientists to measure minute variations in the Moon's gravity, revealing its internal structure in detail for the first time. At the end of the mission in 2012, both spacecraft were crashed into the lunar surface.

▲ A FALSE-COLOR *image showing variations in the Moon's gravity associated with ancient impact craters.*

MOON MISSIONS

The new era of lunar exploration included not just the US, but a number of different nations, including the European Space Agency (ESA), Japan, China, and India.

SELENE

■ **ESA's SMART-1** (2003: orbiter) investigated the theory that the Moon was formed when a smaller planet collided with Earth 4.5 billion years ago.

■ **Japan's Kaguya** (SELENE) (2007: orbiter) released two satellites, Okina and Ouna, into Moon orbit, which helped it map the gravity of the far side of the Moon.

■ **China's Chang'e-1** (2007: orbiter) spent 494 days orbiting the Moon, creating a 3D map of its surface.

■ **India's Chandrayaan-1** (2008: orbiter) searched for radioactive matter that would help researchers explain the Moon's history.

■ **China's Chang'e-2** (2010) orbited the Moon looking for locations for China's next lander, Chang'e-3.

■ **China's Chang'e-4** (2019: lander and rover) made the first successful landing on the far side of the Moon.

■ **India's Chandrayaan-2** (2019) attempted to deploy a lander and a rover near the Moon's south pole.

INTO THE FUTURE

The space nations have several plans for future lunar exploration.

■ **Chang'e-5**—a Chinese mission to collect samples from the lunar surface and return them to Earth.

■ **SLIM**—the first Japanese mission to explore the lunar surface.

■ **Luna-25 & 27**—two planned landers, part of Russia's Luna-Glob program.

Model of a Luna-Glob lander

China on the Moon

Launched in 2013, China's third uncrewed lunar mission, Chang'e-3, made the first soft landing on the Moon since the 1970s. The lander carried cameras, a telescope, and a soil probe and also released a small robot rover called Yutu (meaning "Jade Rabbit") to survey the area around the landing site (👁 p.87). Yutu carried a ground-penetrating radar and two spectrometers—chemical "sniffers" that could analyze minerals in the lunar rocks.

Chang'e-3 lunar lander

A lunar space station

After investigating the idea of a crewed moonbase in the early 2000s, NASA is now planning a space station in orbit around the Moon. Known as the Lunar Orbital Platform-Gateway (LOP-G), this small station will offer easy access to the Moon's surface, as well as providing a low-gravity base where large multipart spacecraft for missions to Mars and beyond can be assembled. NASA hopes to have the station fully assembled and operational by 2026.

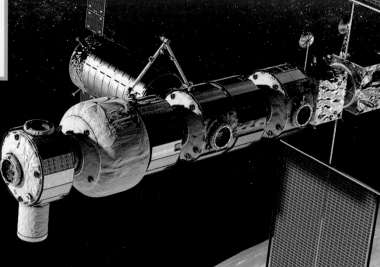

▶ THE LOP-G SPACE STATION *will operate automatically for long periods in between crewed missions of a few weeks at a time.*

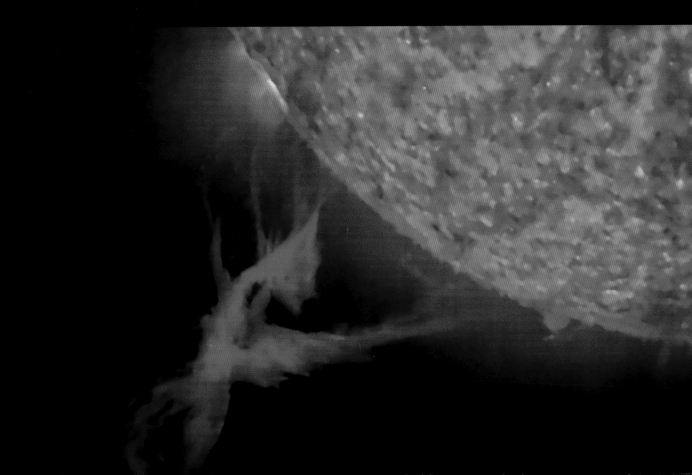

THE SUN

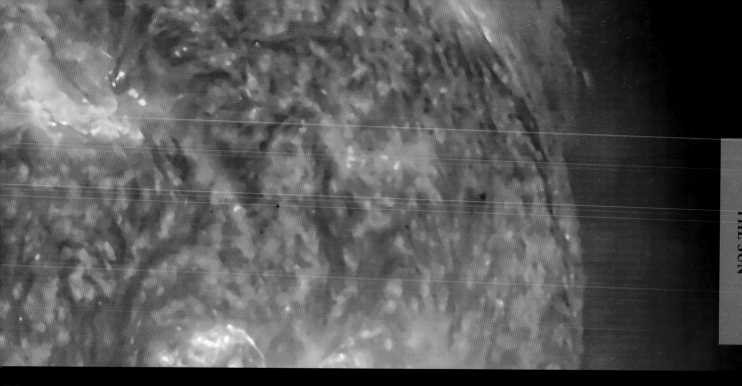

The central star of our solar system is a huge ball of glowing gas 93 million miles (150 million km) away from us. It generates huge amounts of energy inside its core.

The Sun

The Sun is our nearest star, located about 93 million miles (150 million km) from Earth. Even though it's made entirely of gas, its mass is 333,000 times greater than that of Earth and 750 times greater than that of all the planets in the solar system put together.

WATCH THIS SPACE

NASA's Solar Dynamics Observatory (SDO) is one among a fleet of satellites that keep watch over the Sun. It has been studying the Sun at various visible and ultraviolet wavelengths since its launch in 2010. The SDO not only helps us to learn more about the Sun's activity, but also provides forecasts of potentially disruptive solar storms heading for Earth.

▶ SUNSPOTS *are cooler regions of the photosphere, which appear dark against their brighter, hotter surroundings.*

The chromosphere is the layer of atmosphere above the photosphere.

The photosphere is the Sun's visible surface.

The convective zone, through which energy passes in swirls of heated plasma.

SPICULES *are spikes or jets of super-hot plasma forced up through the Sun's magnetic field.*

CORONAL MASS EJECTIONS *are huge bubbles of plasma ejected from the Sun's corona into space.*

THE CORONA *is the uter atmosphere, much otter than the photosphere.*

FACULAE *are hotter, brighter areas of the photosphere that are associated with the formation of sunspots.*

RANULATION *is the ttling caused by convection s at the Sun's surface.*

PROMINENCES *are dense clouds of plasma looping out from the Sun along lines in the magnetic field.*

TELL ME MORE ...

The Sun is fueled by nuclear reactions, which take place within the core. During these reactions, atoms are broken down and huge amounts of energy are released. Temperatures in the core reach 27 million °F (15 million °C). The Sun has been shining for more than 4.6 billion years yet is still less than halfway through its life. Despite burning off around half a billion tons of hydrogen every second, it is big enough to continue shining for at least another 5 billion years.

BIRTH AND DEATH OF THE SUN

Like all stars, the Sun was born in a cloud of gas and dust. About 4.6 billion years ago, the cloud collapsed and gravity broke it up into smaller, denser blobs. These grew hotter and hotter until nuclear reactions started and all the new stars in the cloud began to shine. The Sun will continue to get hotter until it runs out of hydrogen. When this happens, the Sun will grow into a red giant, swallowing the planet Mercury. Finally, the dying star will become a white dwarf, shrouded in a glowing cloud called a planetary nebula.

Inside the Sun

The Sun is a gigantic nuclear power plant. Vast amounts of energy are generated in its superhot core. This filters up to the surface and is emitted into space—mainly in the form of visible light and heat. It is this energy that prevents Earth from turning into a ball of ice.

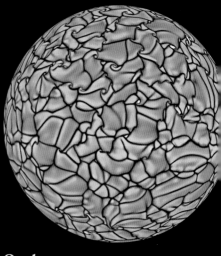

NUCLEAR POWER

The Sun is mainly made of hydrogen gas. Within the core, the crushing pressures and superhot temperatures force hydrogen atoms together. They undergo nuclear fusion and are converted into helium. This process releases huge amounts of energy, which leaves the core in the form of high-energy X-rays and gamma rays.

On the move
Hot gas rising toward the surface from deep inside the Sun creates a pattern o. bright cells. These granulations measur 600–1,200 miles (1,000–2,000 km) across. Larger plumes of rising gas crea giant cells called supergranules, which can measure 18,500 miles (30,000 km. across. Individual granules may last fo up to 20 minutes, while supergranules may last for a couple of days.

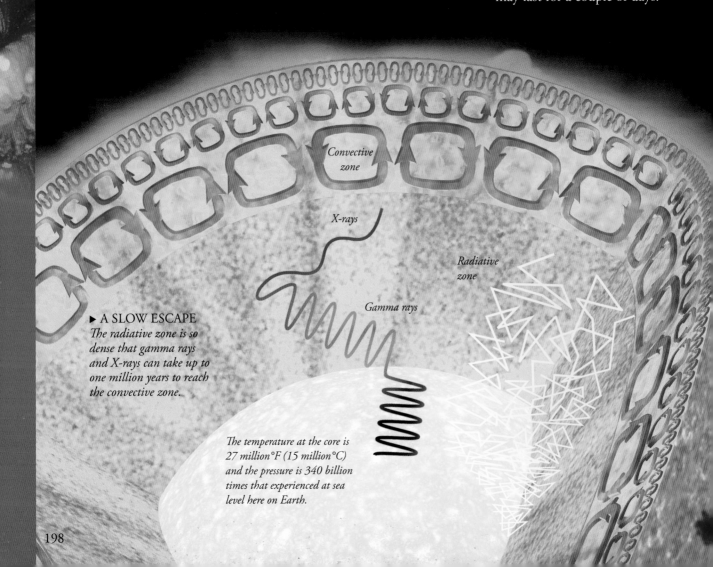

Convective zone

X-rays

Radiative zone

Gamma rays

▶ A SLOW ESCAPE
The radiative zone is so dense that gamma rays and X-rays can take up to one million years to reach the convective zone.

The temperature at the core is 27 million°F (15 million°C) and the pressure is 340 billion times that experienced at sea level here on Earth.

Photosphere

The photosphere is the layer above the convective zone. It is the visible surface of the Sun. The photosphere looks solid but is actually a layer of gas around 300 miles (500 km) thick. It is thin enough to allow light and heat energy to escape out into space. The temperature of this layer is much lower than at the core, around 10,000°F (5,500°C). Light from the photosphere takes about eight minutes to reach Earth.

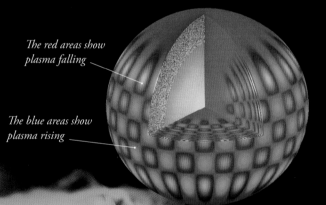

▲ HOT SPOT *The Sun's magnetic field sometimes creates loops of superhot plasma. These pass through the cooler photosphere and shoot up into the corona.*

Burning bright

The Sun releases enough energy per second to meet the needs of Earth's population for more than 1,000 years. It does this by changing 600 million tons (550 million tonnes) of hydrogen into helium every second!

The red areas show plasma falling

The blue areas show plasma rising

The sounds of the Sun

The churning of hot plasma in the convective zone causes sound waves, which travel out through the Sun. At the Sun's surface the waves push the plasma up to 30 miles (50 km) outward, but sound cannot travel through the vacuum of space (which is why we can't hear any noise it makes). Instead, the waves turn inward and allow the plasma to sink back down. By studying these wave patterns scientists have learned a lot about the inside of the Sun.

 TAKE A LOOK: CIRCULATION

The Sun spins about an axis. Unlike Earth, which is solid and has a single speed of rotation, the Sun has several speeds of rotation and spins faster at the equator than at the poles. The surface rotation is illustrated on the right, with the faster areas in green and slower areas in blue. The hot plasma also circulates within the Sun, moving between the equator and the poles. Plasma flowing toward the poles moves fairly close to the surface, but the flow returning to the equator is deeper.

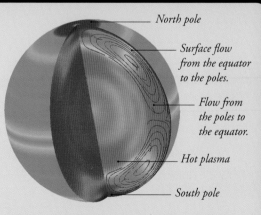

North pole

Surface flow from the equator to the poles.

Flow from the poles to the equator.

Hot plasma

South pole

▲ SOLAR MOTION *The faster areas are shown in green, and the slower areas in blue.*

199

The Sun's atmosphere

The Sun is a huge ball of hot gas. What we see as the surface is the photosphere, the lowest zone of the layered atmosphere, which produces visible light. Above this sit the thin chromosphere and the thick, uneven corona. Each layer is hotter and less dense than the one below it.

THE CORONA

The Sun is surrounded by an extremely hot, wispy atmosphere called the corona. The temperature of the gas here can reach up to 3.6 million °F (2 million °C). Although it is extremely hot, it is not very bright and is usually only seen on Earth during a solar eclipse. However, instruments on spacecraft can now block out the Sun's bright disk so that the corona is visible. The reason for the corona's sizzling temperature is still uncertain, but it seems to be linked to the release of stored magnetic energy.

▲ SOLAR ECLIPSE *The corona appears as a glowing crown around the Moon during a solar eclipse.*

▲ *This image from NASA's TRACE satellite shows the plasma erupting in loops within the corona.*

Coronal loops

Coronal loops are flows of trapped plasma (super-heated gas) that move along channels in the magnetic field of the corona. The plasma flows at up to 200,000 mph (320,000 kph) in loops that can rise more than 600,000 miles (1 million km) above the Sun's surface. They show a wide range of temperatures, and many will reach several million degrees.

▼ AT LEAST TWO *solar eclipses are visible each year to people on Earth. A total solar eclipse, when the Moon completely covers the Sun, can last for several minutes. This is the only time most humans get to view the Sun's outer atmosphere.*

Ulysses

The Sun's poles are very difficult to observe from Earth. To find out more about them, NASA and ESA developed the Ulysses spacecraft. Launched in October 1990, Ulysses is the only spacecraft to have explored the Sun's polar regions. It completed three passes before being shut down in 2009, and revealed that the solar wind is weaker at times of low solar activity.

Dish antenna for communicating with Earth, one of four antennas on Ulysses.

Solar filaments

Huge tongues or arches of relatively cool, dense gas often lift off from the chromosphere and into the corona. They may travel out for hundreds of thousands of miles, sometimes separating from the Sun and launching billions of tons of gas into space. When seen against the brilliant solar disk, they appear as dark ribbons (filaments) but are easily visible as prominences against the blackness of space. Shaped by the Sun's magnetic field, they are often linked to sunspots and solar flares. Some will last for many months, while others will last for only a few hours.

TAKE A LOOK: SOLAR WIND

The Sun releases hot, charged gas particles in a solar wind that blows through space. Particles that escape through holes in the corona create a fast solar wind that blows toward Earth at speeds of up to 560 miles a second (900 km a second). Other areas on the Sun release a solar wind that travels more slowly. These overlapping streams of slow- and fast-moving particles create a shock wave when they meet Earth's magnetic field. Some of the solar wind particles move through this shock wave, passing through the magnetic field and down toward Earth's poles, where they cause the glowing auroras (pp.204–205).

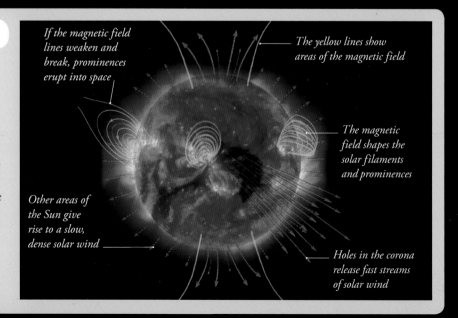

If the magnetic field lines weaken and break, prominences erupt into space

The yellow lines show areas of the magnetic field

The magnetic field shapes the solar filaments and prominences

Other areas of the Sun give rise to a slow, dense solar wind

Holes in the corona release fast streams of solar wind

Solar storms

Breakdowns in the Sun's magnetic field result in violent explosions, which can disable satellites and threaten the lives of astronauts in space. When these eruptions head toward Earth, they can cause dramatic effects in our atmosphere and severe disruptions to our communication systems.

FLARES

Solar flares are huge explosions of high energy radiation that occur around sunspots, where the magnetic field is very intense. They last for only a few minutes but release enormous amounts of energy. Flares may erupt several times a day when the Sun is very active but are rare when the Sun has few sunspots. Major flares can trigger coronal mass ejections.

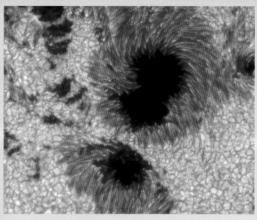

Sunspot cycle
The Sun spins faster at the equator and slower near the poles. This tangles the Sun's magnetic field until, like an overstretched rubber band, it eventually snaps. The field flips and the poles switch around. This event occurs roughly every 11 years and drives the sunspot cycle, a regular rise and fall in the number of sunspots seen on the Sun.

The heat of a solar flare can exceed 18 million °F (10 million °C).

▲ SOLAR POWERED *Solar flares are the biggest explosions in the solar system. They release 10 million times more energy than a volcanic explosion here on Earth.*

TAKE A LOOK: QUAKES

When flares explode they cause quakes inside the Sun, very similar to the earthquakes we experience on Earth. Shock waves from the quake can travel the equivalent of 10 Earth diameters before fading into the photosphere; they can each speeds of up to 250,000 mph (400,000 km/h).

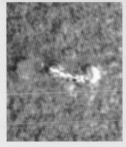

▲ SOLAR FLARE *photographed by the SOHO spacecraft.*

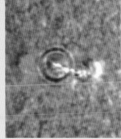

▲ SHOCK WAVES *caused by the flare can be seen in rings around the epicenter.*

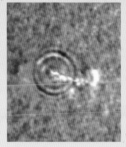

▲ SPREAD *The rings spread out over 60,000 miles (100,000 km) across the Sun's surface.*

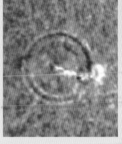

▲ ENERGY *released by the solar quake was huge—enough to power the US for 20 years.*

Mass ejections

Sunspots are often linked with enormous eruptions of gas that blast billions of tons of material out into the solar system. These huge streamers of gas are called coronal mass ejections (CMEs). They fire electrically charged particles out into space at speeds of up to 750 miles per second (1,200 kilometers per second). Reaching the Earth within two to three days, these particles can cause polar auroras, power cuts, and communication disruptions. Like flares, coronal mass ejections are thought to be caused by the rapid release of magnetic energy and are most common at times of peak sunspot activity.

This image shows the largest solar flare ever recorded, observed by SOHO on April 2, 2001.

The flare triggered this massive coronal mass ejection.

WATCH THIS SPACE

In 2001, a magnetic storm raged around planet Earth. Triggered by a coronal mass ejection associated with a giant sunspot, the storm caused spectacular displays of the aurora australis. In the early hours of April 1, the skies over New Zealand were alive with southern lights. Pictured here, the red aurora hangs above the city of Dunedin.

Particle blitz

Charged particles blasted into space by a solar flare blitzed the Solar and Heliospheric Observatory (SOHO) spacecraft only three minutes after the flare erupted on July 14, 2000. The particles created a snowstorm effect on this image taken by the satellite. You can also see a coronal mass ejection blasting a huge cloud of gas into space and the dark circle at the center where the camera blocked the brilliant light from the Sun.

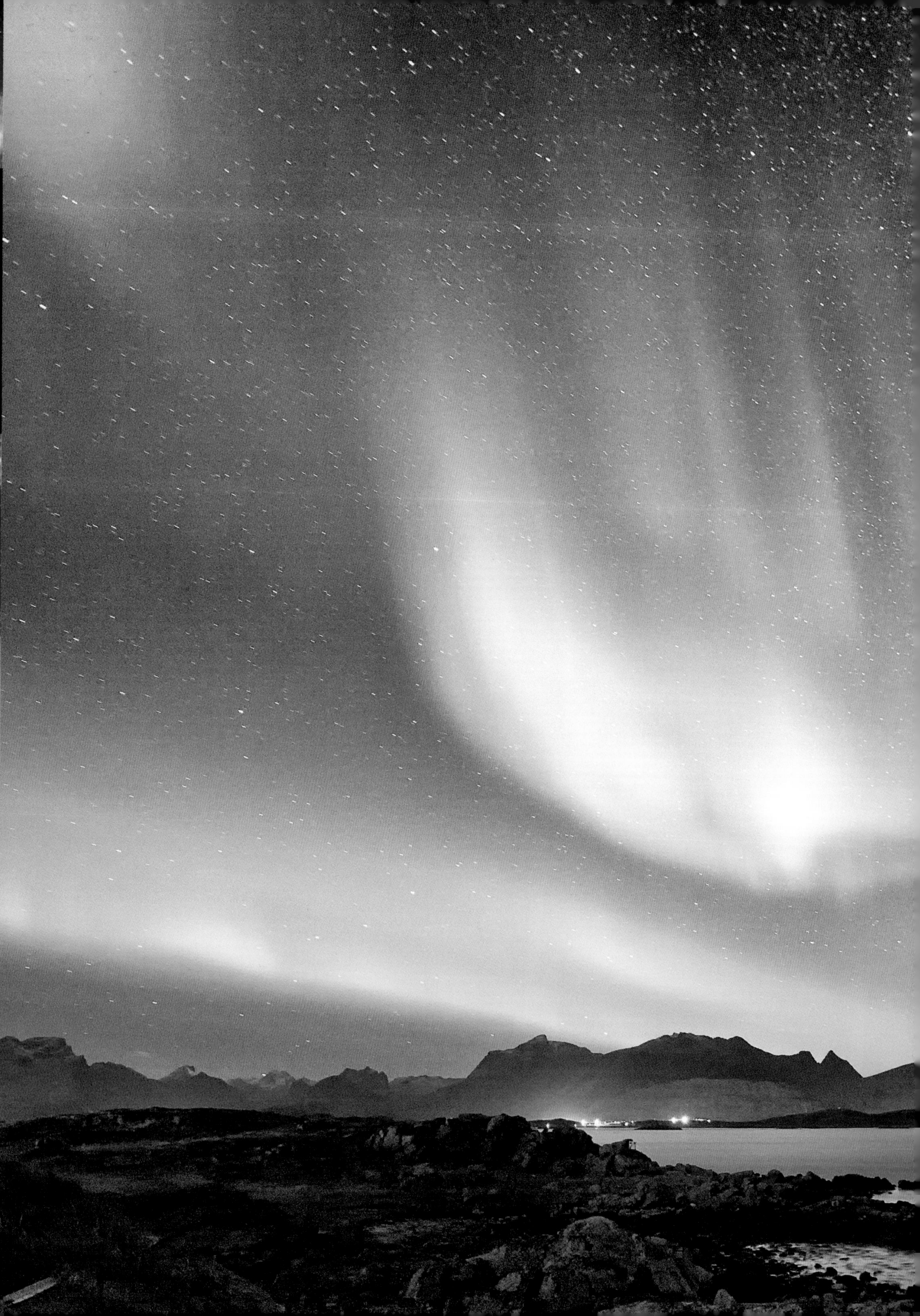

AMAZING AURORA

Auroras are the dancing curtains of light that hang in the polar night sky. They happen when charged particles in the solar wind are trapped by Earth's magnetic field and accelerated onto the upper atmosphere around the magnetic poles. Here, they collide with oxygen and nitrogen atoms, creating green and red glows that take the form of arcs, bands, sheets, and rays.

The solar cycle

Every day, our Sun shines in the sky. Although it always looks the same, it is changing constantly. It goes through cycles of being extremely active followed by periods of quiet. These cycles can have a great effect on our planet.

THE CHANGING SUN

Solar activity varies in a cycle of roughly 11 years. Around the start of each cycle, the Sun's surface is calm and uniform, while peaks of activity see the formation of large sunspot groups and huge solar flares. Despite appearances, the Sun's overall output of radiation only varies by about 0.1 percent over a typical cycle.

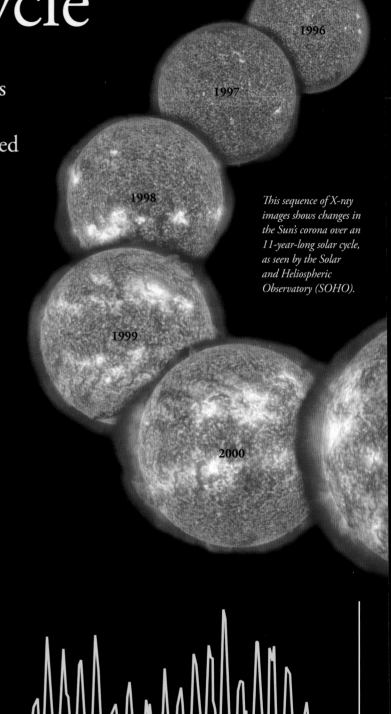

This sequence of X-ray images shows changes in the Sun's corona over an 11-year-long solar cycle, as seen by the Solar and Heliospheric Observatory (SOHO).

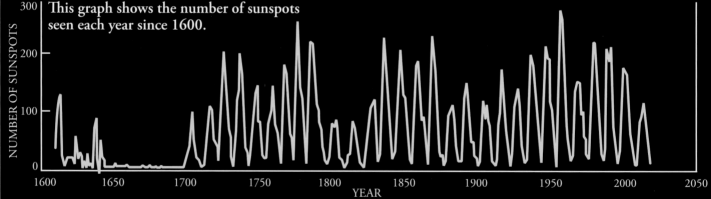

This graph shows the number of sunspots seen each year since 1600.

NUMBER OF SUNSPOTS

YEAR

During the 17th century, almost no sunspots were observed. This period, known as the Maunder Minimum, is the longest recorded period of low solar activity. The Maunder Minimum coincided with a long period of cold weather on Earth, referred to as the "Little Ice Age." Scientists strongly suspect there is a link between the two events.

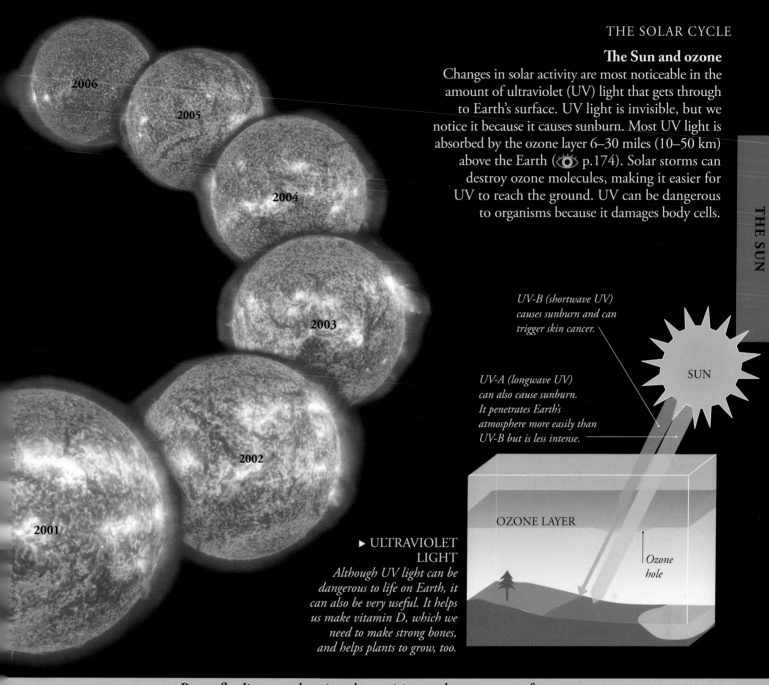

The Sun and ozone

Changes in solar activity are most noticeable in the amount of ultraviolet (UV) light that gets through to Earth's surface. UV light is invisible, but we notice it because it causes sunburn. Most UV light is absorbed by the ozone layer 6–30 miles (10–50 km) above the Earth (👁 p.174). Solar storms can destroy ozone molecules, making it easier for UV to reach the ground. UV can be dangerous to organisms because it damages body cells.

UV-B (shortwave UV) causes sunburn and can trigger skin cancer.

UV-A (longwave UV) can also cause sunburn. It penetrates Earth's atmosphere more easily than UV-B but is less intense.

SUN

OZONE LAYER

Ozone hole

▶ ULTRAVIOLET LIGHT
Although UV light can be dangerous to life on Earth, it can also be very useful. It helps us make vitamin D, which we need to make strong bones, and helps plants to grow, too.

Butterfly diagram showing the position and occurrence of sunspots

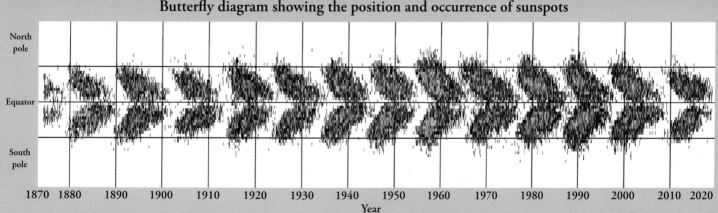

North pole

Equator

South pole

1870 1880 1890 1900 1910 1920 1930 1940 1950 1960 1970 1980 1990 2000 2010 2020
Year

The butterfly effect

English astronomer Edward Walter Maunder (1851–1928) discovered that sunspots do not occur at random over the surface of the Sun. Instead, they follow an 11-year cycle. At the start of each cycle, sunspots appear near the poles, but as the cycle progresses, they appear closer to the equator. When plotting a graph of the sunspot positions he had observed over many years, Maunder realized that the data revealed a butterfly shape—so diagrams of sunspot locations are known as "butterfly diagrams."

Observing the Sun

People have been watching the Sun for thousands of years, keeping records that are used by modern astronomers to understand more about solar activity and past movements of the Sun, Earth, and Moon. Today the Sun is observed by many amateur astronomers and by special solar observatories on Earth and in space.

WHAT A STAR!

The "father of modern astronomy," Italian astronomer Galileo Galilei (1564–1642) proved that the Sun is at the center of the solar system.

GALILEO'S SUNSPOTS

Galileo Galilei studied the Sun by projecting its image through a telescope and drawing what he saw. Making his observations at the same time each day, he noted dark spots on the Sun's surface, which had very irregular shapes and would appear and disappear from the Sun's disk. The movement of the spots also proved that the Sun was rotating on an axis.

Tower telescopes

Close to the ground, heat from the Sun makes the air hot and turbulent. This can distort images received through telescopes, so special tower telescopes are built to observe the Sun. The Richard B. Dunn Solar Telescope at Sacramento Peak in California (right) has a very tall tower, rising 136 ft (41.5 m) above ground level, with another 220 ft (67 m) below ground. Almost all of the air has been removed from the tower to get the clearest possible image of the Sun.

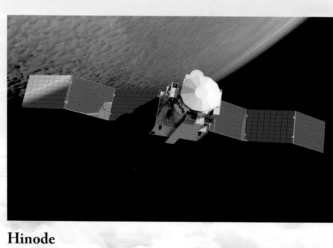

Hinode

Launched in September 2006, the Hinode spacecraft is an orbiting solar observatory created to study the Sun's magnetic activity. It orbits the Earth at an altitude of 370 miles (600 km) and points continuously toward the Sun for 9 months of the year. The spacecraft carries three advanced telescopes, which it uses to take X-ray images of the Sun, to measure its magnetic field in 3D, and to measure the speed of the solar wind.

TAKE A LOOK: THIRTEEN TOWERS OF CHANKILLO, PERU

Located in Peru's coastal desert lies the oldest solar observatory in the Americas. Dating back 2,300 years, the Thirteen Towers of Chankillo are a line of 13 stone blocks running from north to south along a low ridge, forming a "toothed" horizon. The positions of the towers match the points at which the Sun rises and sets over the course of a year. It is likely that the hilltop structure was used as a solar calendar by an ancient Sun cult, helping them to observe the movements of the Sun through the solar year.

June solstice (shortest day)

Equinox

December solstice (longest day)

Observation point

▲ SOLAR CALENDAR *The Thirteen Towers stand like teeth along the ridge, greeting the first and last of the Sun's rays each day.*

Daniel K Inouye Solar Telescope

The largest solar telescope in the world is located at Haleakala Observatory on the Hawaiian island of Maui. With a main mirror 14 ft (4.24 m) in diameter, the Daniel K Inouye Solar Telescope has an advanced computer system to cancel out the blurring of sunlight traveling through the atmosphere. As a result, the telescope can see details on the solar surface as small as 12.5 miles (20 km) across.

STARS AND STARGAZING

  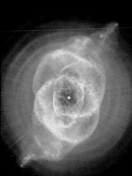

Stars are bright, burning balls of gas that are found all over the universe. They form patterns in our night sky that have been studied for thousands of years.

What are stars?

The Sun, our nearest star, is only 93 million miles (150 million km) away. In terms of the size of the Universe, it's on our doorstep! But the Sun is just one star—there are trillions of others, all with their own amazing features. The Sun is very average in size and brightness and is enjoying a comfortable middle age. But, like all stars, it will change dramatically as it gets older.

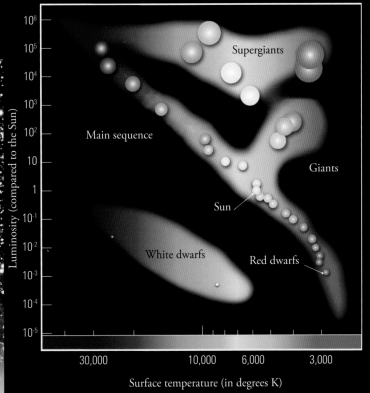

Supergiants

Main sequence

Giants

Sun

White dwarfs

Red dwarfs

Luminosity (compared to the Sun)

10^6
10^5
10^4
10^3
10^2
10
1
10^{-1}
10^{-2}
10^{-3}
10^{-4}
10^{-5}

30,000   10,000   6,000   3,000

Surface temperature (in degrees K)

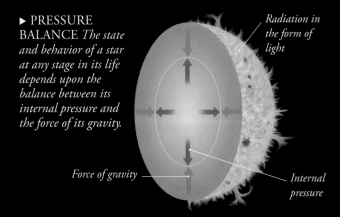

▶ PRESSURE BALANCE *The state and behavior of a star at any stage in its life depends upon the balance between its internal pressure and the force of its gravity.*

Radiation in the form of light

Force of gravity

Internal pressure

HOT AND BRIGHT

This chart (left), called a Hertzsprung-Russell diagram, shows the temperatures of stars and their brightness, or luminosity. Cool stars are shown in red and hot stars in blue. Most hydrogen-burning stars, including our Sun, lie on the diagonal branch, or "main sequence." Giants that have burned all their fuel leave the main sequence, while faint dwarfs lie near the bottom.

THE LIFE OF A STAR

All stars begin life in a cloud of dust and hydrogen gas, called a nebula. Most average stars take billions of years to burn all their hydrogen fuel. When it runs out, the star expands and becomes a red giant, then sheds its outer layers to end its life as a small, dim white dwarf. Bright, massive stars use up their fuel quickly—in a few million years. When there is nothing left to burn, the star expands to become a red supergiant, then explodes in a supernova to form a neutron star or black hole.

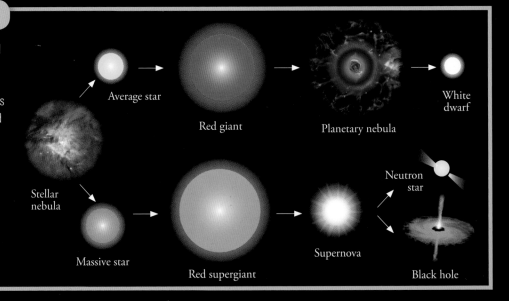

Average star

Red giant

Planetary nebula

White dwarf

Stellar nebula

Massive star

Red supergiant

Supernova

Neutron star

Black hole

STARS AND STARGAZING

TAKE A LOOK: STAR TYPES

Here are some of the types of star found on the Hertzsprung-Russell diagram. All of them are at different stages in their life cycle. Some are young and hot, some are old and cold, and others are about to explode.

◄ WOLF-RAYET STAR *These are very hot, massive stars that are losing mass rapidly and heading toward a supernova explosion.*

▼ MAIN SEQUENCE STAR *Stars like our Sun that lie along the main sequence on the diagram burn hydrogen and turn it into helium.*

▼ BLUE SUPERGIANTS *are the hottest and brightest "ordinary" stars in the Universe. This is Rigel, the brightest star in Orion.*

▲ WHITE DWARF *This is the final stage in the life of an average star like our Sun. A white dwarf is formed from the collapsed core of a red giant and is very dense.*

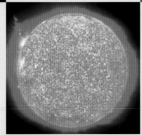

▲ NEUTRON STAR *Formed when a red supergiant explodes, a neutron star is small but extremely dense. Its iron crust surrounds a sea of neutrons.*

▲ RED SUPERGIANT *These stars are huge, with a radius 200 to 800 times that of the Sun, but their surface temperature is low, making them look red or orange-yellow in color.*

Giants and supergiants

When main sequence stars start to run out of fuel, they expand and can become truly enormous. These giant and supergiant stars swell up and start to burn helium instead of hydrogen. One day, our own Sun will turn into a red giant about 30 times bigger and 1,000 times brighter than it is today.

VV Cephei

Antares

Betelgeuse

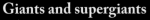

Sun (1 pixel) Sirius Pollux Arcturus Aldebaran Rigel

▲ TRUE GIANT *Even supergiants like Betelgeuse and Antares are dwarfed by VV Cephei, a star so big that it is known as a hypergiant. It lies in the constellation Cepheus, about 2,400 light-years from Earth, and is one of the largest-known stars in the Milky Way.*

Birth of a star

Most stars are born in a huge cloud of gas and dust, called a nebula. The story starts when the nebula begins to shrink, then divides into smaller, swirling clumps. As each clump continues to collapse, the material in it becomes hotter and hotter. When it reaches about 18 million °F (10 million °C), nuclear reactions start and a new star is made.

NEBULAS

Nebulas can be different colors. The color comes from the dust in the nebula, which can either absorb or reflect the radiation from newborn stars. In a blue nebula, light is reflected by small dust particles. A red nebula is caused by stars heating the dust and gas.

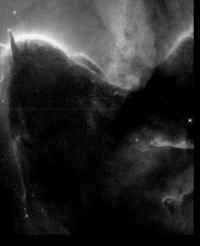

▲ THE TRIFID NEBULA
This cloud of gas and dust lies in the constellation of Sagittarius. The cloud is gradually being eroded by a nearby massive star. At the top right of the cloud, a stellar jet is blasting out from a star buried inside. Jets like these are the exhaust gases from newly forming stars.

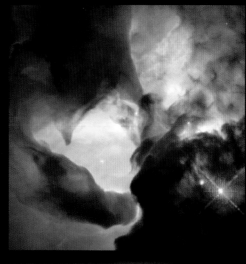

▲ THE LAGOON NEBULA
Near the Trifid is the even larger Lagoon Nebula. It gets its name from a dark patch that looks like a lake. Several groups of new stars are forming inside this nebula. At its center is a very young, hot star whose radiation is evaporating and blowing away the surrounding clouds.

▲ THE HORSEHEAD NEBULA

Not all nebulas are colorful. The black Horsehead Nebula is a cloud of cold dust and gas in the constellation Orion. The horse's head shows up against the red nebula behind it, which is heated by stars. Many stars have formed in the region of Orion within the last million years.

▲ THE SEVEN SISTERS

The Pleiades cluster lies in the constellation of Taurus. It is also known as the Seven Sisters, because up to seven of its massive, white-hot stars can be seen with the naked eye. There are more than 300 young stars in the cluster, surrounded by a thin dust cloud that shows as a pale blue haze.

▲ THE EAGLE NEBULA

This huge finger of cool hydrogen gas and dust forms part of the Eagle Nebula. At the top of this finger, hot, young stars shine brightly among the dark dust. Eventually, these stars will blow the dust away and become clearly visible as a new star cluster.

○ TAKE A LOOK: CARINA NEBULA

These two images show part of the Carina Nebula—a huge pillar of dust and gas where stars are being born. In the top image, the cloud is glowing due to radiation from nearby stars. The infrared image (bottom) allows us to see some of the stars inside the nebula.

▲ VISIBLE LIGHT *Hidden inside this glowing nebula are stars that have yet to emerge.*

▲ INFRARED LIGHT *Here, two infant stars inside the nebula are releasing jets of material.*

A FLASH OF BRILLIANCE

V838 Monocerotis is a red supergiant star, located about 20,000 light-years away from Earth. In March 2002, this star suddenly flared to 10,000 times its normal brightness. The series of images below shows how a burst of light from the star spread out into space, reflecting off the layers of dust that surround the star. This effect is called a light echo. The images make it look as if the nebula itself is growing, but it isn't. The spectacular effect is caused by light from the stellar flash sweeping outward and lighting up more of the nebula.

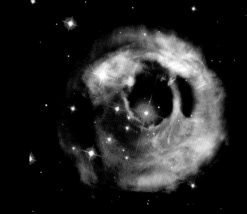

May 20, 2002

September 2, 2002

October 28, 2002

December 17, 2002

September 2006
More than four years after the star erupted, the echo of the light is still spreading out through the dust cloud.

The death of a star

BEFORE *This star is about to explode.*

10 DAYS AFTER *This image shows the same star during its supernova explosion. The star is situated in a nearby galaxy called the Large Magellanic Cloud. When it exploded in 1987, it was the first supernova to be visible to the naked eye for almost 400 years.*

The larger a star is, the shorter its life will be. Hot, massive stars shine only for a few million years because they burn up their hydrogen fuel rapidly. Smaller stars are much cooler, so they use their fuel more slowly and can shine for billions of years. But, sooner or later, all stars run out of fuel and die.

Betelgeuse

When a star begins to use up its hydrogen fuel, it balloons outward to become a huge red giant or supergiant. Betelgeuse, a red supergiant in the constellation of Orion, is more than 1,000 times wider than the Sun. It is also about 14,000 times brighter, because it is burning its fuel at a rate 14,000 times faster than the Sun. A few hundred thousand years from now, Betelgeuse will have exhausted its fuel and will explode as a supernova. It will then become the brightest star in our sky, second only to the Sun.

Stellar death throes

Eta Carinae is a star that is rapidly reaching the end of its life. It is being torn apart by massive explosions that throw out huge clouds of gas and dust. The star's brightness is also changing dramatically. In 1843, it was the second brightest star in the sky; today, it cannot be seen with the naked eye.

SMOKE RINGS

Small- or medium-sized stars like our Sun end up as red giants. When a red giant runs out of hydrogen and helium, it is not hot enough to burn other fuels, so it collapses. Its outer layers are puffed out into space like giant smoke rings. These shells of gas are called planetary nebulas, because they looked like planets when first seen through early telescopes. The central star shrinks to form a white dwarf, an extremely hot object about the size of Earth.

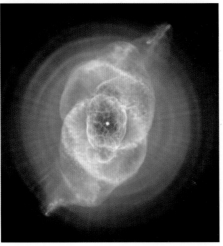

▲ THE CAT'S EYE NEBULA
The central bubble of gas was ejected by the dying red giant star about 1,000 years ago. It is expanding outward into older gas clouds created by previous outbursts.

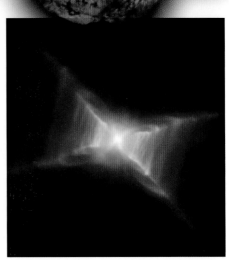

▲ THE RED RECTANGLE NEBULA
At the center of this nebula is a binary (double) star system. The two stars are surrounded by a ring of thick dust, which has shaped the surrounding gas into four spikes.

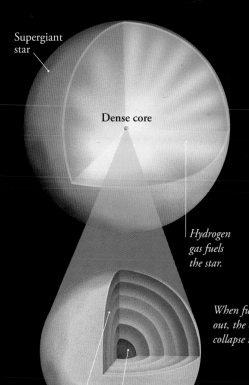

Supergiant star

Dense core

Hydrogen gas fuels the star.

Other heavy elements

Innermost core made of iron

SUPERNOVA

Big stars, with a mass at least eight times the mass of our Sun, die in a spectacular way. As they run out of fuel, they suddenly collapse, then the outer layers of the star are blasted outward in a huge explosion known as a supernova. The energy released by a supernova is as much as the energy radiated by the Sun during its entire lifetime. One supernova can outshine a galaxy containing billions of stars. Supernovas are rare events—none have been seen in our galaxy since the invention of the telescope. The nearest supernova of recent times occurred in the Large Magellanic Cloud Galaxy in February 1987.

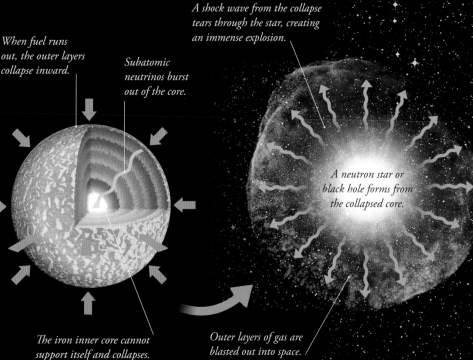

When fuel runs out, the outer layers collapse inward.

Subatomic neutrinos burst out of the core.

A shock wave from the collapse tears through the star, creating an immense explosion.

A neutron star or black hole forms from the collapsed core.

COLLAPSE OF A STAR *A supernova is caused by a star collapsing and then exploding. All that remains of the star after the explosion is a black hole or a dense neutron star surrounded by an expanding cloud of gas.*

The iron inner core cannot support itself and collapses.

Outer layers of gas are blasted out into space.

▲ THE EGG NEBULA
Here, the central star is hidden by a dense layer of gas and dust. However, its light illuminates the outer layers of gas, creating a series of bright arcs and circles.

▲ THE BUTTERFLY NEBULA
This nebula consists of two "wings" of gas thrown out from the dying central star. The butterfly stretches for about 2 light-years— half the distance from our Sun to the next star.

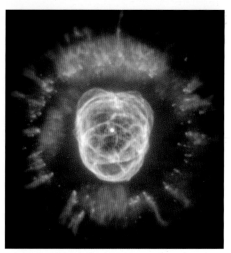

▲ THE ESKIMO NEBULA
The "parka hood" is a ring of comet-shaped objects, with their tails streaming away from the star. The "face" is a bubble of material being blown into space by the star's wind.

Interstellar space

The space between the stars, called interstellar space, is not completely empty—there are scattered molecules of gas and dust everywhere. Over a whole galaxy, this adds up to a huge amount of material.

GLOBULES

Small clouds of gas and dust are called globules. The smallest are known as Bok globules, after the US astronomer Bart Bok, and are often as small as our solar system (about 2 light-years across). The gas in these clouds is mainly molecular hydrogen, with a temperature of around –436°F (–260°C). Globules can contract slowly under the force of their own gravity and form stars.

▶ BOK GLOBULES
These dark Bok globules are silhouetted against a background of hot, glowing hydrogen gas.

Gas and dust

Scientists can detect molecules in space because they absorb or emit radio waves. Nearly 200 types of molecules have been identified so far. The most common are gases, such as hydrogen. There is enough gas in the Milky Way, for example, to make 20 billion stars like our Sun. Dust particles, water, ammonia, and carbon-based (organic) compounds have also been found in space.

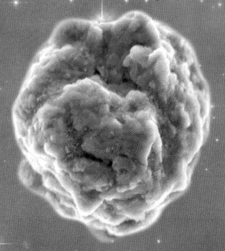

▲ SPACE DUST *Each dust particle is smaller than the width of a human hair.*

Globule with a tail

Looking like an alien monster about to swallow a helpless galaxy, this faint, glowing cloud of dust and gas is being shaped by winds from a nearby, newly born star. The star's strong ultraviolet light makes the cloud's "mouth" glow red. This cloud is an example of a cometary globule, so called because its long tail resembles the tail of a comet.

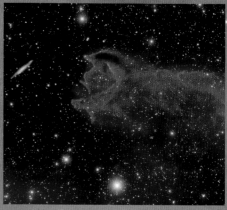

▲ DUST CLOUD *This globule contains enough material to make several stars as big as our Sun.*

▼ THE VEIL NEBULA
lies in the constellation of Cygnus, the Swan.

THE VEIL NEBULA

Dust and gas are continually being added to interstellar space by stellar winds and dying stars. The wispy Veil Nebula is the remains of a massive supernova that exploded about 5,000 years ago. Even today, the Veil Nebula is still expanding outward at a rate of about 60 miles (100 km) each second.

Birthplace of stars

This cloud, called the Orion Nebula, is so bright that it can easily be seen with the naked eye. It lies about 1,500 light-years from Earth, measures about 25 to 30 light-years across, and has a mass several hundred times that of the Sun. The Orion Nebula is heated by a group of young stars at its center, called the Trapezium, and is a place where new stars are being formed.

Solar system on the move

Our solar system is sweeping through interstellar space at high speed. As it moves, the solar wind creates an invisible bubble around it known as the heliosphere. This bubble pushes against the gas and dust in interstellar space, forcing the gas and dust to flow around it. Scientists once thought that the shape of the solar system as it moved through space was like a comet with a tail, but new observations show that it actually resembles a squishy ball.

> ## ▶▶▶ FAST FACTS ▶▶▶
>
> ■ The ingredients of interstellar space are continually changing as new molecules are created and others are split apart.
> ■ Dust and gas are added by dying stars and removed by the birth of new stars.
> ■ Hydrogen, helium, and carbon monoxide are the most common gases in space.
> ■ Space is bathed with many forms of radiation, such as light, heat, and radio waves.
> ■ Other space ingredients include magnetic fields, cosmic rays, and neutrons.

◀ THE TRAPEZIUM
The cluster around the Trapezium contains 1,000 hot stars that are less than a million years old.

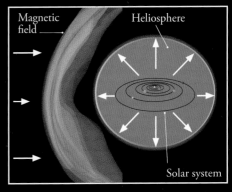

Magnetic field   Heliosphere

Solar system

▲ PUSHING THROUGH SPACE *The interstellar magnetic field bends and parts to let the solar system pass through.*

Multiple stars

Most stars form in clusters inside huge clouds of gas and dust. As time goes by, these stars may drift apart until they are no longer part of the original cluster. Our Sun is fairly rare in being a solitary star. More than half of all stars are in binary systems, while many others are in systems of three or more stars.

BINARY SYSTEMS

A binary system is a group of two companion stars that orbit each other. The first binary to be discovered was Mizar, situated in the "handle" of the Big Dipper. Its companion star was spotted by Giovanni Riccioli in 1650. Since then, many pairs of double stars have been found. Famous binaries include the bright star Acrux in the Southern Cross, which was discovered to be double in 1685, and Mira, a red giant in Cetus (the Whale).

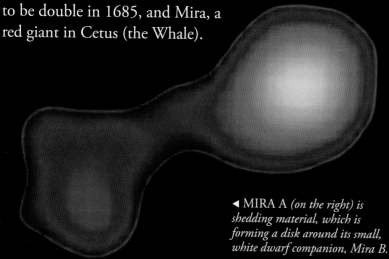

◀ MIRA A *(on the right) is shedding material, which is forming a disk around its small, white dwarf companion, Mira B.*

Double Dog Star

The brightest star in the night sky is Sirius, nicknamed the Dog Star because it is in the constellation of Canis Major (the Great Dog). The blue-white star Sirius A is hotter than our Sun and 22 times as bright. Its companion, Sirius B, is a faint white dwarf, the dense remnant of a collapsed star.

▲ SIRIUS B *(on the right) is so close to Sirius A and so faint that images of it have only recently been obtained.*

Cannibals in space

Sometimes the two stars in a binary system are so close that one is able to steal material from the other. The "cannibal" star then grows in size and mass at the expense of its neighbor. One example of this is the double star system Phi Persei. This contains an elderly star that is shedding its outer layers. The cast-off material has been sucked in by its companion, which has now grown to a hefty nine times the size of our Sun. It is spinning so violently that it is flinging gas from its surface into a ring around its middle. One day, it may even start dumping gas back onto the first star.

THE PHI PERSEI DUO

1. THE PAIR OF STARS *in Phi Persei have stayed the same for the last 10 million years, orbiting one another and held together by the pull of their gravity.*

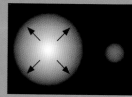

2. THINGS CHANGE *when the bigger star starts to run out of hydrogen— the fuel that powers its nuclear furnace. The now-aging star begins to swell.*

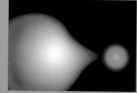

3. AS THE AGING STAR *expands, it begins to dump its mass onto its smaller companion star.*

4. THE FIRST STAR *sheds practically all of its mass, leaving its bright core exposed.*

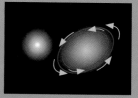

5. THE SMALLER COMPANION *has now captured most of its partner's excess mass. It changes identity from a moderately sized star into a massive, hot, rapidly spinning star.*

6. THE SECOND STAR *is spinning so fast that its shape is distorted into a flattened sphere. The spinning also causes the star to shed hydrogen gas, which settles into a broad ring around it.*

OPEN CLUSTERS

Open clusters are groups of hundreds or even thousands of stars. They are held together by their gravity, which attracts them to one another. The stars in an open cluster all formed inside the same large cloud of gas and dust. As a result, they are all the same age and have the same composition, but their masses can vary considerably. Well-known open clusters that are visible with the naked eye include the Pleiades (The Seven Sisters), the Hyades, and the Jewel Box.

Three's a crowd

There is more to the Pole Star (Polaris) than meets the eye—it is actually a triple star. One companion, Polaris B, has been known since 1780. The third star is so close to Polaris A that it wasn't seen until 2005.

▲ NGC 3603 *This giant nebula contains one of the biggest clusters of young stars in the Milky Way Galaxy. This image shows young stars surrounded by dust and gas.*

Globular clusters

Dense, ball-shaped groups of stars, called globular clusters, orbit the Milky Way and other large galaxies. A single cluster can contain millions of stars, which all formed at the same time and from the same cloud. These stars can stay linked by their gravity for billions of years. Many globulars are very old and contain some of the oldest surviving stars in the Universe.

 FAST FACTS

■ The age of most globular clusters suggests that they formed very early in the history of the Universe, when the first galaxies were being born.

■ Most globulars are full of elderly stars, typically 10 billion years old, and no new stars are forming.

■ However, some globular clusters contain several generations of younger stars, so they must have formed more recently.

■ Young globular clusters may be the leftovers of collisions between large galaxies and dwarf galaxies.

▲ REMAINS OF A DWARF GALAXY?
Omega Centauri is one of the most spectacular sights in the southern night sky. This globular cluster is thought to be around 12 billion years old. Recent observations show that stars near its center are moving very rapidly, suggesting that the cluster has a medium-sized black hole at its center. The cluster may be the old heart of a dwarf galaxy that was largely destroyed in an encounter with the Milky Way.

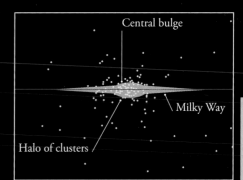

Central bulge

Milky Way

Halo of clusters

▲ GLOBULARS CLOSE TO HOME

There are over 150 globular clusters near the Milky Way. Unlike open clusters, which are always found in the disk of the Milky Way Galaxy, many globular clusters are located in a "halo" around the galaxy's central bulge. Scientists can calculate how far away these globulars are from how bright they appear.

▲ WHITE AND RED DWARFS

NGC 6397 is one of the closest globular star clusters to Earth. The Hubble Space Telescope has been able to look right into the center of this cluster. It found faint white dwarfs that died long ago, as well as faint, cool red dwarfs that have been slowly burning up their hydrogen fuel for possibly 12 billion years.

Mega cluster

Omega Centauri is the biggest of all the Milky Way's globular clusters, containing perhaps 10 million stars and measuring about 150 light-years across. In the night sky, it appears nearly as large as the full Moon.

▲ M13 *This globular cluster is one of the brightest and best known in the northern sky. The glittering ball of stars appears to the naked eye as one hazy star and is easily spotted in winter in the constellation Hercules. About 300,000 stars are crowded near its center, with more scattered farther out. M13 measures more than 100 light-years across.*

Other solar systems

For centuries, people have wondered whether distant stars had planets orbiting around them. Unfortunately, most stars are so far away that it was impossible to spot any planets. But modern instruments have now made it possible to detect them, and thousands have already been found.

BABY PLANETARY SYSTEMS

Out in space, new solar systems are still forming. This is the Orion Nebula, where many stars are being born. Around each new star is a spinning disk of gas and dust. If material in this disk starts to clump together, it eventually forms planets that orbit the star.

Exoplanets

A planet situated outside our solar system is called an exoplanet. The first two were discovered in 1992 in orbit around an extreme type of star called a pulsar. These planets cannot be seen, but their existence is known from the way they affect the radio waves emitted by the pulsar (p.227).

▲ PULSAR PLANETS *These planets are unlikely to support life, because pulsars emit high levels of harmful radiation.*

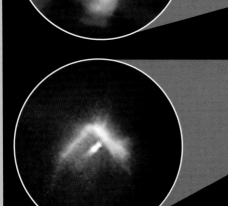

▲ PLANETARY NURSERY *Astronomers have found 30 baby solar systems forming in the Orion Nebula.*

Pulling power

The first exoplanet in orbit around a Sun-like star was discovered in 1995. The planet was detected from a tiny wobble in the motion of the star 51 Pegasi. As the planet, called 51 Pegasi b, orbited the star, its gravity sometimes pulled the star toward Earth and sometimes away from it. This wobble showed up as slight shifts in the spectrum of the starlight. Since then, hundreds of exoplanets have been found from the wobbles they create in nearby stars.

▲ COLOR SHIFTS *The wavelength of a star's light changes as it moves toward or away from Earth. Shifts in the spectrum may show that a planet is present.*

Unseen planet

DIPPING LIGHT

Another way of finding exoplanets is to look for changes in the brightness of a star when an orbiting planet passes in front of it and blocks some of its light.

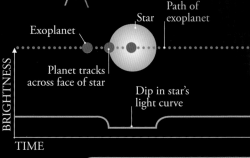

Path of exoplanet

Star

Exoplanet

Planet tracks across face of star

Dip in star's light curve

BRIGHTNESS

TIME

PROXIMA B is the closest exoplanet to our solar system. Discovered in 2016 from shifts in the light of Proxima Centauri, a small red star 4.2 light-years away, it could be a rocky world (as shown in this artist's impression) just a little larger than Earth.

DUSTY DISKS

Planets form inside huge rotating disks of dust and gas. Even before the first exoplanets were spotted, dust disks were found around many young stars. The first was the disk around a star called Beta Pictoris. In 2008, scientists discovered an object very close to this star. They think it is a giant planet orbiting every 20 years or so.

◀ BETA PICTORIS *is a hot young star in the constellation Pictor. The disk around the star is quite cool but glows brightly in infrared light.*

55 CANCRI

Among the alien solar systems discovered so far, the 55 Cancri system, in the constellation Cancer, is one of the most like our own. Our solar system has eight planets, while 55 Cancri has at least five—more than any other exoplanet system that has so far been discovered. The inner four planets of 55 Cancri are all closer to the star than Earth is to the Sun, and all five of its planets are larger than Earth. Both systems have a giant gas planet in a distant Jupiter-like orbit. However, this planet lies in the habitable zone for the star, and liquid water could exist on a rocky moon orbiting it.

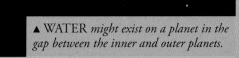

▲ WATER *might exist on a planet in the gap between the inner and outer planets.*

227

Extreme stars

The Universe is full of stars that are hotter, colder, more massive, or less massive than our Sun. Some of these extreme stars are at the end of their lives. Some are stars that have suddenly become very active. Others are failed stars that never ignited their nuclear furnaces.

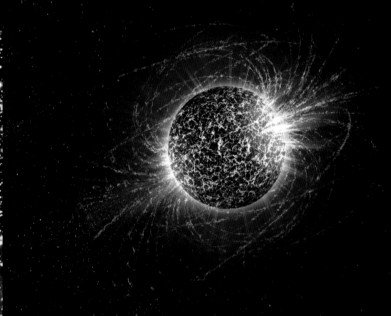

NEUTRON STARS

Neutron stars are small, only about 6 miles (10 km) across, yet they are heavier than the Sun. One teaspoon of material from a neutron star would weigh a billion tons. Neutron stars are covered by an iron crust 10 billion times stronger than steel. Inside, they contain a liquid sea of neutrons— the debris from atoms crushed by a supernova explosion.

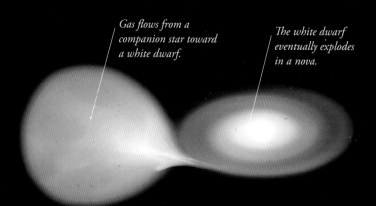

Gas flows from a companion star toward a white dwarf.

The white dwarf eventually explodes in a nova.

DWARF STARS

White dwarfs

Any star with a mass less than seven times our Sun is expected to end its life as a small, dim star known as a white dwarf. When a dying star puffs off most of its material and collapses, it becomes extremely small, dense, and hot. The matter in a white dwarf is so densely packed that a teaspoonful of the material would weigh several tons.

White dwarf

▲ END OF A STAR *Our Sun will end its life as a white dwarf, like these stars, about 7 billion years from now.*

Brown dwarfs

Some stars, known as brown dwarfs, are so small and cool that they are unable to start up nuclear reactions in their core or to burn hydrogen. They are often described as "failed stars." Brown dwarfs do shine, but very faintly, because they produce a little heat as they slowly shrink due to gravity.

▲ TWIN BROWN DWARFS *This artwork shows the dimmest starlike bodies known, called 2M 0939.*

Novas

If a white dwarf orbits close to a normal star in a binary system, it can pull large amounts of gas from the other star. This gas gets extremely hot, pressure increases on the white dwarf's surface, and eventually a huge nuclear explosion occurs. The white dwarf then grows dimmer for a period of weeks or months before the same thing happens again. These periodic explosions are called novas.

PULSARS

A pulsar is a neutron star that emits pulses of radiation as it rotates. When seen from Earth, these pulses appear to sweep across the night sky like the beam from a lighthouse. The radiation from a pulsar can be experienced on Earth as radio signals or sometimes flashes of visible light, X-rays, and gamma rays.

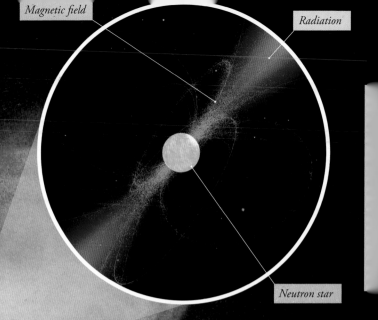

Magnetic field

Radiation

Neutron star

▲ PULSES OF RADIATION
A neutron star has an intense magnetic field and rotates rapidly, producing high-energy electrons that radiate into space.

Extreme outbursts

Observatories sometimes detect powerful but short-lived bursts of gamma rays. These flashes are brighter than a billion Suns yet last only a few milliseconds. They are thought to be caused by a collision either between a black hole and a neutron star or between two neutron stars. In the first case, the black hole drags in the neutron star and grows bigger, as shown below. In the second type of impact, the two neutron stars create a black hole.

◀ STELLAR QUAKES
In 2004, one magnetar flared up so brightly that it temporarily blinded all the X-ray satellites in space. The blast of energy came from a giant flare created by the star's twisting magnetic field.

MAGNETARS

Magnetars are a type of neutron star with magnetic fields up to 1,000 times stronger than those of other neutron stars. They are the strongest known magnets in the Universe, equal to 10 million million hand magnets. Their intense magnetism may result from them spinning very quickly—300 to 500 times a second—when they are born. This spin, combined with churning neutron fluid in the interior, builds up an enormous magnetic field.

Black holes

A black hole is possibly the strangest object in the Universe. It is a region of space where matter has collapsed in on itself. This results in a huge amount of mass being concentrated in a very small area. The gravitational pull of a black hole is so strong that nothing can escape it—not even light.

▲ BIG AND SMALL *Black holes come in many sizes. Some are only a few times more massive than our Sun. Others, found at the centers of galaxies, may be millions or billions of times more massive. This photograph shows a black hole at the center of galaxy M87 which is about 6.5 billion times more massive than the Sun. It is the first time that we've spotted a black hole.*

Stellar mass black holes

This type of black hole forms when a heavyweight star—about 10 times heavier than our Sun—ends its life in a supernova explosion. What is left of the star collapses into an area only a few miles across. A stellar mass black hole is most easily found when it has a companion star that survives the explosion. Material is often pulled off this star and forms a disk swirling around the black hole. Experts can then calculate the black hole's mass and orbit.

Disk of hot material

▲ TO A CREWMATE, *an astronaut looks normal as he starts to be pulled toward the black hole.*

Companion star

▲ JETS OF RADIATION *stream away from the black hole at nearly the speed of light.*

▶ LONG AFTER *the astronaut has fallen into the black hole, crewmates see him, highly stretched and red, on its rim.*

Stretched beyond the limit

Objects that fall into black holes are stretched to just one atom wide. An astronaut who fell in feet first would feel a stronger pull of gravity on his feet than his head. This stretching effect would get worse closer to the hole, and eventually he would be crushed by its overpowering gravity. Crewmates watching from a distance would see him turn red as light struggled to escape from the black hole, appear to hover on the edge of the hole, then disappear.

▲ TWO HOLES *These bright objects are two supermassive black holes orbiting one another. Eventually, they may collide to form one enormous black hole. The pink streaks are the jets that they blast out.*

SUPERMASSIVE BLACK HOLES

Most galaxies, including the Milky Way, are believed to contain supermassive black holes at their centers. Some experts think that these black holes are created when a lot of material is squeezed together in the center of a newly forming galaxy. Another possibility is that supermassive black holes start very small and then grow gradually by pulling in and swallowing nearby material.

Jet of
radiation

▶ COSMIC JETS
As gas is drawn into a black hole, it gets very hot. This energy is released as jets of radiation (usually X-rays) that are blasted far out into space.

Ring of dust and gas

Jet of
radiation

 FAST FACTS

■ All the matter that falls into a black hole piles up at a single point in the center called the singularity.

■ If two black holes collided, they would cause gravity waves to ripple through the whole Universe.

■ To turn Earth into a black hole, it would have to be squashed to the size of a marble!

■ There may be as many as 100 billion supermassive black holes in our part of the Universe alone.

■ Black holes are slowly losing all their energy, but it will take billions of years before they evaporate into nothing.

Be a *skygazer*

People have been fascinated by the night sky since prehistoric times. Early civilizations recorded the positions of the Sun, Moon, and planets. Today, light from street lamps and buildings can hide the fainter stars, but there are still plenty of amazing views for skygazers.

▲ ESSENTIAL EQUIPMENT *As well as a star map, take books with you to find out more about what you are looking at. Use a red light to read by; if you use an ordinary flashlight, it will take longer for your eyes to readjust to the darkness. Finally, don't forget to wear warm clothing!*

SEEING STARS

If you want to see small, faint objects in the sky, pick a good spot away from bright light. Take a moment for your eyes to adjust to the darkness around you, and then use binoculars or a telescope. Binoculars are cheaper than telescopes and are good for looking at star fields, star colors, clusters, and the Moon. Telescopes magnify more and are better for planets, nebulas, and galaxies.

SIGNPOSTS IN THE SKY

At first glance, the night sky seems to be pretty evenly scattered with stars, but if you keep looking, patterns begin to emerge. These star patterns, or constellations, were named by early astronomers. One of the most noticeable constellations is Orion (right). It is one of the best signposts in the northern winter sky and can be used to find other constellations and bright stars.

D = 76mm F=600mm

To Castor and Pollux

To Procyon

To Aldebaran

To Sirius

FOCAL RATIO 1:7.9

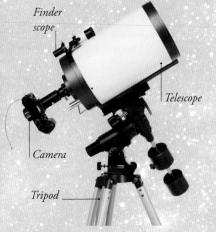

Finder scope

Telescope

Camera

Tripod

▲ LONG SHOT *Pictures of very faint objects can be taken by attaching a camera to a telescope and leaving the shutter open for a few minutes.*

The colorful Universe

The colors of planets and stars can be easy to see, but nebulas and galaxies are often disappointing—even in large telescopes, they look like gray or greenish fuzzy patches. This is because their light is not bright enough for the color-sensing part of your eye to pick up. To see the colors, you need to take pictures of star trails or nebulas with a camera. Hold the camera shutter open for a few minutes while keeping the camera steady.

 TAKE A LOOK: STAR MAPS

The stars in the night sky are so far away that their positions look fixed. You might find it pretty easy to remember where the brightest stars and constellations are, but to find the fainter objects, you'll need a star map. There are different types of these. A paper map is useful but difficult to handle—especially on a breezy night! A planisphere is a disk that you turn to show the exact part of the sky above you. Maps are also available on the internet. Star map apps are useful, too.

Line up the numbers and turn the disk to match the time and date.

The area revealed in the window shows what's in the sky above you.

▲ GUIDING STAR *A planisphere will help you find your way around the stars.*

SUNGAZING

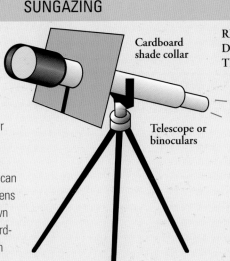

▲ *Looking directly at a solar eclipse can damage your eyes.*

The Sun is fascinating to watch, but it's so bright, it can cause blindness. The safest way to look for sunspots or study a solar eclipse is to project an image of the Sun onto a piece of cardboard. You can use a telescope or one lens of binoculars to shine an image of the Sun onto paper (shown here). You can make a pinhole projector. Cover a piece of cardboard with foil and pierce a tiny hole in it. With this, you can project an image of the Sun onto another cardboard piece.

Cardboard shade collar

Telescope or binoculars

REMEMBER: NEVER LOOK DIRECTLY AT THE SUN, EVEN THROUGH SUNGLASSES.

Paper with magnified image of the Sun

The night sky

If you look up into the sky on a clear night, you will see thousands of stars, but how do you know which star is which? Luckily, the stars form groups known as constellations, which can help you find your way around the night sky.

Southern constellations

Northern constellations

WHO DREW THE CONSTELLATIONS?

Early astronomers noticed that the stars formed groups and that these groups moved in a regular way across the heavens. They began to use characters, animals, and objects from their myths and legends to remember these groups. Most of the constellation names we use today date from Greek and Roman times, but some were invented more recently in the 17th and 18th centuries.

STAR CATALOGS

Early Western astronomers drew up catalogs of the constellations. At first, only 48 constellations were known because much of the southern hemisphere had not been explored by Europeans, so the southern constellations had not been seen. As sailors began to venture farther south, more and more constellations were added. In 1922, the International Astronomical Union accepted an official set of 88 constellations we know today and defined their shapes. However, some groups around the world still use their own constellations.

▶ ASTRONOMERS *looking through a telescope, as illustrated on a page in* Harmonia Macrocosmica, *a 17th-century star atlas by cartographer Andreas Cellarius.*

TAKE A LOOK: PLANETS

Venus

Moon

Stars are not the only things that are visible in the night sky—you can also spot planets. Mercury, Venus, Mars, Jupiter, and Saturn are all visible to the naked eye. Mercury and Venus are known as the morning and evening stars, because the best times to see them are just before sunrise or after sunset.

Finding the Pole Star

The Pole Star sits almost directly above the North Pole, which makes it an excellent way to find due north. It is visible all year in the northern hemisphere at the tip of a constellation called Ursa Minor (the Little Bear). To find it, you can use another constellation called Ursa Major (the Great Bear). Seven of its stars form a shape that is known as the Big Dipper or the Plough. The two stars that form the front of this shape point to the Pole Star, which is the next bright star you see.

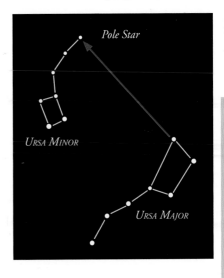

Pole Star

URSA MINOR

URSA MAJOR

THE ZODIAC

A group of 12 constellations can be seen in both hemispheres. Ancient Greek astronomers called them the zodiac, from the Greek word for animals. Most of them are named after animals, but some are human and one is an object. The zodiac runs along a path in the sky called the ecliptic, which is at an angle of 23½ degrees to the equator. The Sun, Moon, and planets also move on paths close to the ecliptic.

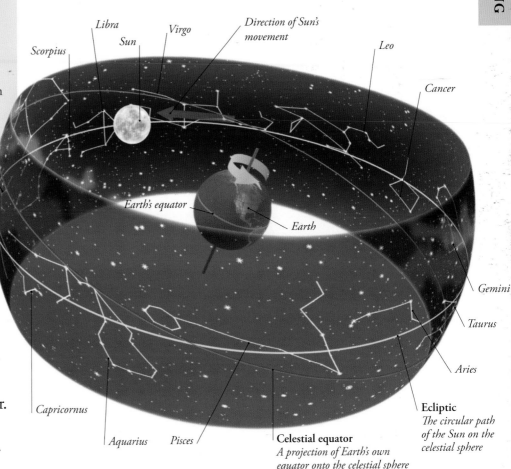

Libra

Scorpius

Sun

Virgo

Direction of Sun's movement

Leo

Cancer

Sagittarius

Earth's equator

Earth

Gemini

Taurus

Aries

Capricornus

Ecliptic
The circular path of the Sun on the celestial sphere

Aquarius

Pisces

Celestial equator
A projection of Earth's own equator onto the celestial sphere

CONSTELLATIONS ON THE MOVE

The stars that we see in a constellation look as if they are grouped together, but in fact, some are much closer to us than others. They appear to be flat against the sky because our eyes can't determine the distances between them. Each star is also moving in space. In a few hundreds of thousands of years' time, the stars will all be in different positions and the constellations will have changed shape from how we know them today.

Big Dipper 100,000 years ago …

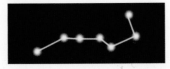

as it is today …

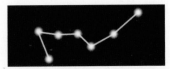

and in 100,000 years' time.

The *northern* sky

To spot constellations, you need a star chart and a place with a wide view of the sky. The chart on the right shows the constellations visible from the northern hemisphere. You will not be able to see all of them at once—the Earth's tilt and orbital motion mean that some can only be seen at certain times of year.

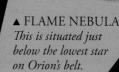
▲ FLAME NEBULA
This is situated just below the lowest star on Orion's belt.

▼ ORION NEBULA *(M42) is a huge area of star formation situated in the "sword" that hangs from Orion's belt.*

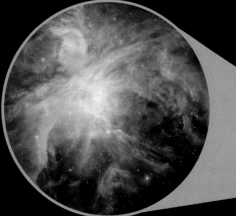

Betelgeuse

Rigel

Star names
Some stars, such as Betelgeuse and Rigel in the constellation Orion, were named in ancient times. But most stars are identified by letters and numbers. For the brightest stars, astronomers use Greek letters and the constellation name. In this system, Betelgeuse and Rigel are Alpha Orionis and Beta Orionis, respectively.

Orion
The Hunter

Orion is one of the most easily recognizable constellations in both the northern and southern skies. It depicts a hunter armed with a club and a sword that hangs from the three diagonal stars making up his belt. He is holding the head of a lion. The Orion constellation contains two very bright stars—**Rigel**, a blue supergiant at the bottom right, and **Betelgeuse**, a red supergiant at the top left.

Orion

Cygnus
The Swan

Cygnus is a major constellation of the northern hemisphere, sometimes called the **Northern Cross.** It can also be seen close to the horizon in the southern hemisphere in winter. At the base of the swan's tail is the bright star **Deneb**, a blue-white supergiant 160,000 times brighter than the Sun. The beak of the swan contains a double star, **Albireo**, whose two stars can be seen with binoculars or a small telescope.

Cygnus

▼ USING THE CHART *Turn the book until the current month is in front of you. You may find it easier to photocopy the page, stick it onto some cardboard, and cut it out. Then face south and look for the stars as they appear on the map. If you are not sure which direction is south, make a note of where the Sun is at noon. That direction is always south.*

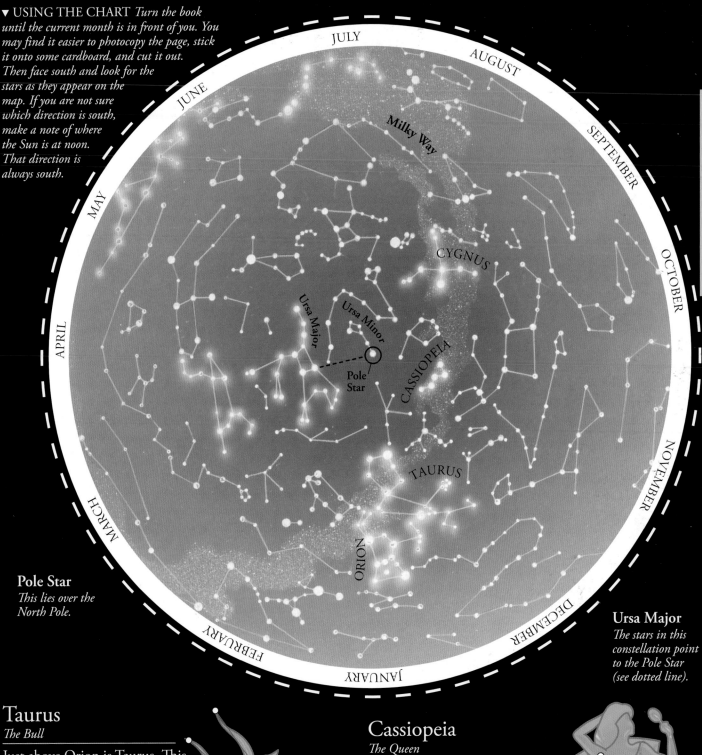

JULY

AUGUST

SEPTEMBER

OCTOBER

NOVEMBER

DECEMBER

JANUARY

FEBRUARY

MARCH

APRIL

MAY

JUNE

Milky Way

CYGNUS

Ursa Major

Ursa Minor

CASSIOPEIA

Pole Star

TAURUS

ORION

Pole Star
This lies over the North Pole.

Ursa Major
The stars in this constellation point to the Pole Star (see dotted line).

Taurus
The Bull

Just above Orion is Taurus. This constellation features two famous star clusters called the **Hyades** and the **Pleiades**, both of which contain stars that are visible with the naked eye. A prominent red star called Aldebaran forms the eye of the bull, while just above the star that marks the tip of the bull's lower horn is the Crab Nebula (M1). This supernova remnant is all that remains of an exploding star, first spotted in 1054.

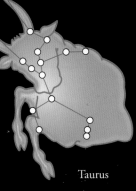

Taurus

Cassiopeia
The Queen

Another easily recognizable constellation is Cassiopeia. It is named after a mythical queen who was notoriously vain, which is why she is shown with a mirror in her hand. The five main stars in this constellation form a distinctive W shape. The center star of the W points toward the Pole Star.

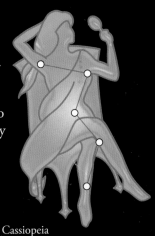

Cassiopeia

The *southern* sky

Stargazing in the southern hemisphere can give you views you may not catch in the northern hemisphere, such as looking into our galaxy's center or finding our closest neighbor star system, Alpha Centauri. Here are some of the most interesting things to look for.

▲ TRIFID NEBULA *This colorful nebula is divided into three lobes and contains some very young, hot stars.*

Milky Way

▲ LAGOON NEBULA *This huge nebula, visible with the naked eye, appears pink in images taken through telescopes.*

CENTER OF THE GALAXY

When we look at the night sky, we can see other parts of our galaxy, the Milky Way. It is at its most dense in the constellation of Sagittarius, because here we are looking right into the center of the galaxy. Sagittarius contains more star clusters and nebulas than any other constellation.

Sagittarius
The Archer

Sagittarius is depicted as a centaur, a mythical half-man, half-horse creature, firing an arrow. This constellation contains a radio source, thought to be a black hole, which marks the center of the **Milky Way Galaxy**. Sagittarius also contains the **Lagoon**, **Trifid**, and **Omega** nebulas and the globular cluster **M22**.

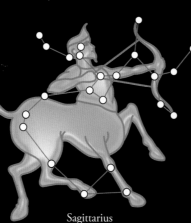

Sagittarius

Hydra
The Water Snake

The biggest of all the 88 constellations, Hydra spreads across nearly a quarter of the sky. Most of the stars it contains are very faint. The brightest star in this constellation is a giant star called **Alphard**. Hydra also contains two star clusters and a planetary nebula.

Hydra

TELL ME MORE …

These websites have more information about the monthly or weekly night sky:

- www.skyandtelescope.com/ interactive-sky-chart/
- www.heavens-above.com /skychart2.aspx

▼ USING THE CHART *Turn the book until the current month is in front of you. Then face north and look for the stars as they appear on the map. If you don't have a compass to find north, make a note of the direction of the Sun at noon, which is always north in the southern hemisphere.*

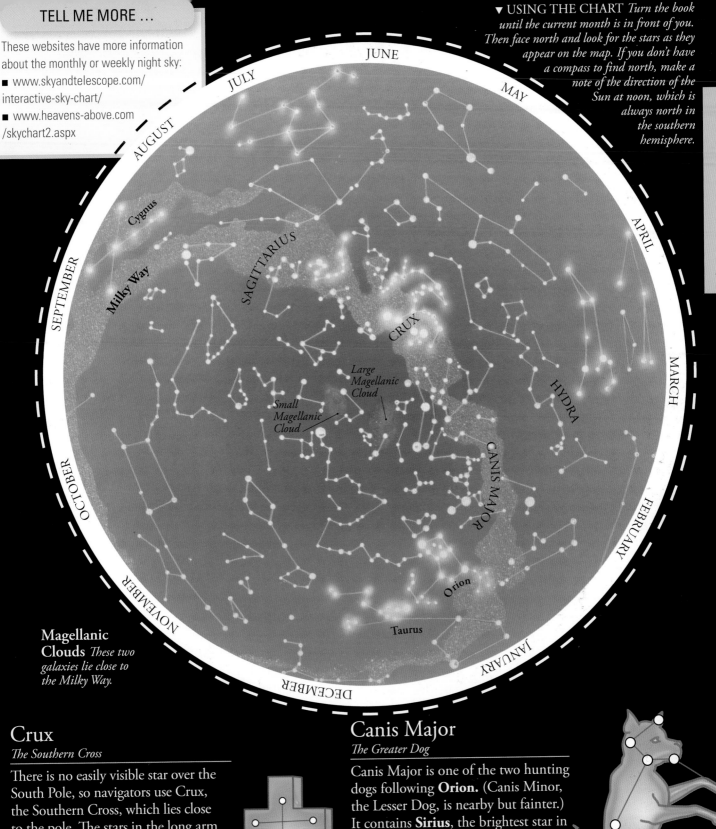

Magellanic Clouds *These two galaxies lie close to the Milky Way.*

Crux
The Southern Cross

There is no easily visible star over the South Pole, so navigators use Crux, the Southern Cross, which lies close to the pole. The stars in the long arm of the cross point toward the pole's position. Although Crux is the smallest of the constellations, it contains four very bright stars, one of which is a red giant. Lying close to the left arm of the cross is the **Jewel Box** cluster of stars, just visible with the naked eye.

Crux

Canis Major
The Greater Dog

Canis Major is one of the two hunting dogs following **Orion.** (Canis Minor, the Lesser Dog, is nearby but fainter.) It contains **Sirius**, the brightest star in the sky, also known as the Dog Star. Sirius has a companion white dwarf star, but this can only be seen with a powerful telescope. Sirius was important in the Egyptian calendar, as it heralded the annual flooding of the Nile and the start of the new year.

Canis Major

Space *in time*

People have been fascinated by the night sky for thousands of years. Observations from astronomers throughout the centuries have greatly expanded our knowledge of how the Universe works.

▼ **1845** *Jean Foucault and Armand Fizeau take the first detailed photographs of the Sun's surface through a telescope—the first space photographs ever taken.*

▼ **1781** *William Herschel discovers Uranus while using one of his telescopes. He first thought it was a comet.*

▼ **1846** *Johann Gottfried Galle identifies Neptune.*

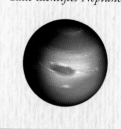

▲ **1609** *Galileo Galilei builds his own telescope to study the stars. His discoveries helped prove that the Sun is at the center of the solar system.*

▲ **2300 BCE** *Stonehenge is built, thought to be a giant stone astronomical calendar.*

| 3000 BCE | 1600 CE | 1700 | 1800 |

▼ **1801** *Giuseppe Piazzi discovers Ceres, the first asteroid. William Herschel is the first to use the term "asteroid" in 1802.*

▲ **164 BCE** *Astronomers from Babylon in the Middle East record the earliest known sighting of Halley's comet. It is seen again in 1066 CE and recorded on the Bayeux Tapestry (above).*

▼ **1655** *Christiaan Huygens observes Saturn and discovers its rings.*

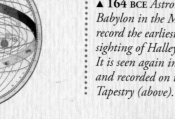

▲ **320–250 BCE** *The Greek astronomer Aristarchus of Samos is the first to suggest that the Earth travels around the Sun. It took 18 centuries before people agreed with this idea.*

▲ **1895** *Konstantin Tsiolkovsky is the first to suggest that rockets can work in a vacuum, making spaceflight possible.*

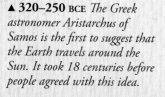

▲ **1916** *German physicist Karl Schwarzschild works out theories that lead to the idea of black holes.*

▼ **1931** *Georges Lemaitre suggests the theory that the Universe started from a single atom. His "cosmic egg" idea later becomes known as the "The Big Bang Theory."*

▲ **1959** *Russia's moon probe Luna 2 is the first spacecraft to land on the Moon, and Luna 3 sends the first photographs of the far side of the Moon back to Earth.*

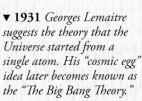

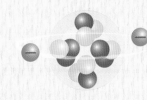

▼ **1926** *The first liquid-fuel rocket is launched by Robert Goddard.*

▲ **1961** *Yuri Gagarin is the first person in space, orbiting Earth for 108 minutes!*

1900 ────────────── **1950** ──

▼ **1957** *Sputnik 1 is launched into orbit by Russia. It is the first human-made satellite in space.*

▼ **1962** *NASA's Mariner 2 is the first space probe to reach a planet as it flies past Venus. This is the start of many more spaceflights by the US and Russia in the 1960s and 1970s.*

▲ **1930** *Subrahmanyan Chandrasekhar predicts the idea of supernovas, caused by large white dwarf stars collapsing in on themselves.*

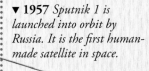

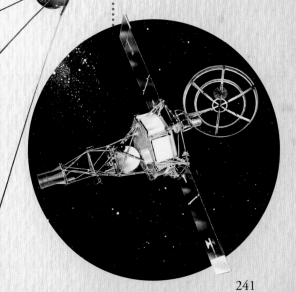

▲ **1945** *Arthur C. Clarke, a science-fiction writer, suggests that a satellite can be used for transmitting telephone and TV signals around Earth. His ideas become reality 20 years later.*

▲ **1925** *Edwin Hubble announces the discovery of galaxies beyond our own.*

◄ **1986** Mir is the first permanent space station in orbit. It enables people to live in space for extended periods of time.

▲ **1965** Russian Alexei Leonov was the first person to spacewalk. He spent 12 minutes floating up to 17½ ft (5 m) from Voskhod 2.

▼ **1981** The first of NASA's reusable space shuttles, Columbia, is flown into space.

▼ **1976** NASA's Viking 1 is the first spacecraft to land on, and explore, Mars.

▲ **1969** Neil Armstrong flies into space on Apollo 11 and is the first person to walk on the Moon.

1970   **1980**

► **1971** Lunokhod 1 finishes its mission as the first remote-controlled lander on the Moon.

▼ **1986** The European Space Agency's Giotto probe takes the first ever close-up photographs of a comet nucleus as it flies through Halley's comet.

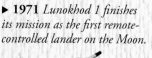

▲ **1977** NASA launches the Voyager probes to explore deep space.

▼ **1971** Russia's Salyut 1, the world's first space station, is launched into orbit.

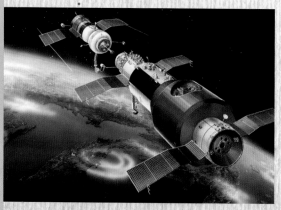

▲ **1982** Rings discovered around Neptune.

▼ **2001** *Genesis probe is launched to collect samples of atoms from the solar wind.*

▼ **1994** *Hubble Space Telescope uncovers evidence of a black hole in the M87 galaxy.*

▲ **2004** *SpaceShipOne is the first privately built spacecraft to reach outer space.*

▼ **2001** *NEAR is the first spacecraft to orbit and land on an asteroid (Eros).*

▼ **2010** *NASA retires the space shuttles. Their last flight was in September 2010.*

1990 **2000**

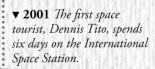

▼ **2001** *The first space tourist, Dennis Tito, spends six days on the International Space Station.*

▲ **2006** *Stardust mission uses aerogel to bring back samples of comet dust.*

THE FUTURE?
There are still many discoveries to be made. The biggest challenges include finding ways to explore farther afield in space and finding life on other planets.

▲ **1990** *The Hubble Space Telescope is the first large optical telescope in orbit. After its mirror is fixed, it returns amazing pictures of distant stars and galaxies.*

▶ **1998** *The first modules of the International Space Station are launched.*

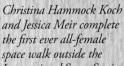

▶ **2019** *US astronauts Christina Hammock Koch and Jessica Meir complete the first ever all-female space walk outside the International Space Station.*

Glossary

Absorption line A dark line or band on a spectrum that corresponds to the absorption of light at a particular wavelength.

Aerogel A lightweight substance used to collect space dust.

Antenna A device used on spacecraft and telescopes to send and receive signals.

Aphelion The point in the orbit of a planet, comet, or asteroid, when it is farthest away from the Sun.

Asteroid A giant rock that orbits the Sun.

Asteroid belt The area of space in the solar system that has the highest number of orbiting asteroids in it, between the orbits of Mars and Jupiter.

Astrolabe An ancient instrument used to calculate the position of stars in the sky.

Astronaut A person trained to travel in a spacecraft.

Atmosphere The layer of gas that surrounds a planet.

Atom The smallest particle of matter that can exist on its own. It is made up of neutrons, protons, and electrons.

Aurora Curtains of light that appear near the poles of planets. Solar wind particles are trapped by a magnetic field and are drawn into the planet's atmosphere. Here they collide with atoms and give off light.

Axis The imaginary line that goes through the center of a planet or star and around which it rotates.

Background radiation A faint radio signal that is given out by the entire sky; leftover radiation from the Big Bang.

Big Bang The cosmic explosion that scientists believe created the universe billions of years ago.

Billion One thousand million.

Binary stars Two stars that orbit each other. It is also called a binary system.

Black hole An area of space with such a strong gravitational pull that it sucks in anything that comes too close, even light.

Blazar An active galaxy that has a supermassive black hole at its center and sends high-speed jets of gas toward Earth.

Brown dwarf An object that is smaller than a star, but larger than a planet. It produces heat, but little to no light at all.

Celestial object Any object that is seen in the sky.

Charged particle A particle that has a positive or negative electrical charge.

Chromosphere The region of the Sun's atmosphere above the photosphere.

Comet A large solid object made of dust and ice that orbits the Sun. As it gets near the Sun, the ice starts to vaporize, creating a tail of dust and gas.

Constellation Originally a pattern of stars but now simply an area of sky defined by astronomers.

Coriolis effect An effect of Earth's rotation that makes winds and ocean currents swirl to the right in the northern hemisphere and to the left in the southern hemisphere.

Corona The Sun's hot upper atmosphere. It is seen as a white halo during a solar eclipse.

Cosmonaut A Russian astronaut.

Cosmos Another word for the universe.

Crater A hollow or basin made by a meteorite crashing into a planet or the Moon.

Crust The thin outer layer of rock of a planet or moon.

Dark energy The energy that scientists believe is responsible for the expansion of the universe.

Dark matter Invisible matter that can bend starlight with its gravity.

Density The amount of matter that occupies a certain volume.

Drag The force that opposes the forward movement of something through the air.

Dust Tiny bits of "soot" from stars that absorb starlight. Also fine material on the surfaces of planets and moons.

Dwarf planet A planet that is big enough to have become spherical but has not managed to clear all the debris from its orbital path.

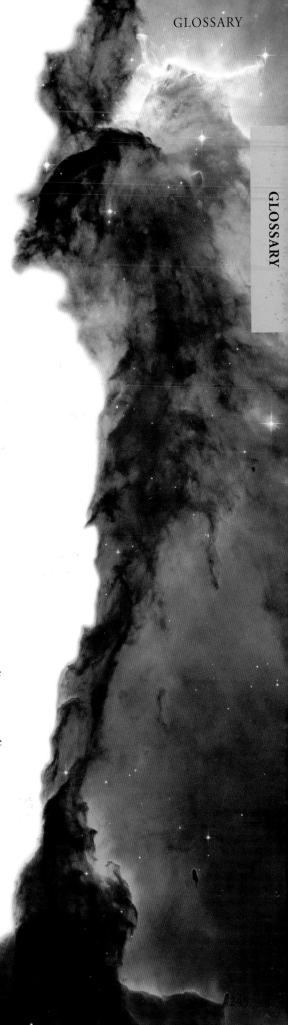

Eclipse The blocking of light from an object when another object passes in front of it. A lunar eclipse is when the shadow of Earth falls on the Moon. A solar eclipse is when the shadow of the Moon falls on the Earth.

Electromagnetic radiation Energy waves that can travel through space and matter.

Electromagnetic spectrum The complete range of energy waves in order of wavelength, from radio waves to gamma rays.

Electron A subatomic particle with a negative electrical charge.

Elliptical Oval-shaped.

Equator The imaginary line around the center of a planet.

Escape velocity The speed at which an object has to travel to escape the gravity of another object.

EVA Short for "extra-vehicular activity," which means activity by an astronaut outside of his or her spacecraft in space.

Exoplanet A planet outside our solar system.

Exosphere The top layer of Earth's atmosphere, where most spacecraft fly.

Extraterrestrial Not belonging to Earth.

False-color image A picture of an object where different colors are used to show up matter or features that we can't normally see in visible light. Images from non-optical telescopes are shown in false color.

Filament A string of galaxy superclusters that stretches out across space. Also name for a huge tongue of gas released into space from the surface of the Sun.

Flyby When a spacecraft flies past a planet, comet, or asteroid without landing or orbiting it.

Free-fall A state of weightlessness that occurs when an object is not affected by gravity, or any opposing force, for example, in orbit around Earth.

Galaxy A collection of millions of stars, gas, and dust held together by gravity and separated from other galaxies by empty space.

Gamma ray An energy wave that has a very short wavelength.

Geostationary orbit The orbit of a satellite that moves around Earth at the same speed as Earth, so that it looks as if it is not moving across the sky.

Geyser A jet of liquid that escapes through cracks in rock.

Globular clusters Ball-shaped groups of stars that orbit large galaxies.

Globules Small clouds of gas and dust in space.

Granulation Mottling on the surface of the Sun.

Gravity The force that pulls objects toward one another.

Habitable If a place is habitable, it is suitable for living in, or on.

Heliopause The boundary between the heliosphere and interstellar space.

Heliosphere A large area that contains the solar system, the solar wind, and the solar magnetic field.

Hemisphere Half of a sphere. The division of Earth into two halves, usually by the equator, which creates a northern hemisphere and a southern hemisphere.

Hertzsprung-Russell diagram A diagram that shows a star's temperature, brightness, size, and color in relation to other stars.

Hydrothermal Relating to heated water inside Earth's crust.

Hypersonic Relating to the speed of something that is equal to or more than five times the speed of sound.

Infrared Waves of heat energy that can't be seen.

Intergalactic Between galaxies.

Interstellar Between the stars.

Ionosphere A region of Earth's atmosphere 30–375 miles (50–600 km) above the surface.

K Stands for degrees Kelvin, a measurement of temperature. 0 Kelvin (absolute zero) is –459°F (–273°C).

Launch vehicle A rocket-powered vehicle that is used to send spacecraft or satellites into space.

Light Waves of energy that we can see.

Light-year The distance that light travels in one year.

Low Earth orbit An orbit close to Earth.

Luminosity The brightness of something.

Magnetar A type of neutron star with an incredibly strong magnetic field.

Magnetic field An area of magnetism created by a planet, star, or galaxy, which surrounds it and extends into space.

Magnetometer An instrument that is used to measure magnetic forces.

Magnetosphere The area around a planet where the magnetic field is strong enough to keep out the solar wind.

Magnitude The brightness of an object in space, shown as a number. Bright objects have low or negative numbers and dim objects have high numbers.

Mantle A thick layer of hot rock underneath the crust of a moon or planet.

Mare A large, flat areas of the Moon that looks dark when viewed from Earth. They were originally thought to be lakes or seas, but are now known to be floods of lava. The plural is maria.

Matter Something that exists as a solid, liquid, or gas.

Mesosphere The layer of atmosphere 30–50 miles (50–80 km) above the Earth, where shooting stars are seen.

Meteor A bit of rock or dust that burns up as it enters the Earth's atmosphere. They are also called "shooting stars."

Meteorite A rocky object that lands on Earth.

Microgravity When the force of gravity is present, but its effect is minimal.

Microwave A type of energy wave with a short wavelength.

Milky Way The name of the galaxy where we live.

Module A portion of a spacecraft.

Multiverse Universes that are parallel to our own.

Nebula A cloud of gas and dust in space from which stars are born.

Neutrino A particle smaller than an atom that is produced by nuclear fusion in stars and by the Big Bang. It is very common, but extremely hard to detect.

Neutron A subatomic particle that does not have an electrical charge.

Neutron star A dense, collapsed star that is mainly made of neutrons.

Nucleus The center of something.

Observatory A building, spacecraft, or satellite containing a telescope that is used for observing objects in space.

Orbit The path an object travels around another object while being affected by its gravity.

Orbiter A spacecraft that is designed to orbit an object, but not land on it.

Ozone Colorless gas that forms a layer in Earth's atmosphere, absorbing some of the harmful ultraviolet radiation in sunlight.

Particle An extremely small part of a solid, liquid, or gas.

Payload Cargo that is carried into space by a launch vehicle or on an artificial satellite.

Perihelion The point in the orbit of a planet, comet, or asteroid, when it is closest to the Sun.

Phase The amount of the Moon or a planet's surface that is seen to be lit up by the Sun.

Photosphere The part of the Sun's lower atmosphere where its light and heat come from.

Planet A celestial object that orbits a star.

Planetary nebula A glowing cloud of gas and plasma around a star at the end of its life.

Planetesimals Small rocky or icy objects that are pulled together by gravity to form planets.

Planisphere A moveable disk that shows the position of the stars in the night sky

Plasma A highly energized form of gas.

Probe An unmanned spacecraft that is designed to explore objects in space and transmit information back to Earth.

Prominence Large flamelike plume of plasma that comes out of the Sun.

Proton A subatomic particle with a positive electrical charge.

Pulsar A neutron star that sends out pulses of radiation as it spins.

Quasars Short for quasi-stellar objects, which means a very luminous, distant object that looks like a star.

Radiation Energy released by an object.

Radiometer A piece of equipment used for detecting or measuring radiation.

Red giant A very bright, but very cool huge star.

Rille A narrow channel or crack on the Moon's surface.

Rover A vehicle that is driven over the surface of a planet or moon, usually by remote control.

Satellite A naturally occurring or man-made object that orbits another object larger than itself.

Seyfert galaxy An active galaxy, often a spiral, powered by a supermassive black hole at its center.

Shock wave A wave of energy that is produced by an explosion or by something traveling at supersonic speed.

Silicate A type of mineral containing silicon and oxygen.

Solar radiation Energy from the Sun.

Solar wind A flow of charged particles from the Sun.

Space–time The combination of all three dimensions of space together with time.

Stratosphere The layer of atmosphere 10–30 miles (16–50 km) above Earth where airplanes fly.

Subatomic particles A particle that is smaller than an atom and that makes up an atom.

Suborbital A type of orbit where a spacecraft flies to the top of Earth's atmosphere (60 miles, 100 km) and weightlessness occurs.

Supernova The bright explosion that occurs as a star collapses.

Taikonaut A Chinese astronaut.

Thermosphere The layer of atmosphere 50–370 miles (80–600 km) above the Earth, where auroras occur.

Thrust The force produced by a jet or rocket engine that pushes something forward.

Transit The passage of a planet or star across the face of another.

Troposphere The layer of Earth's atmosphere 0–10 miles (0–16 km) above the ground, where our weather occurs.

Ultraviolet ray A type of energy wave. It is an important part of sunlight, but exposure to it can burn people's skin.

Umbra The dark, central area of the Moon's shadow or of a sunspot.

White dwarf A small, dim star. Our Sun will eventually become a white dwarf.

X-ray A type of energy wave that can pass through objects that visible light cannot penetrate.

Zero gravity Not in fact a lack of gravity, but an apparent lack of gravity experienced by astronauts in free-fall or in orbit.

Index

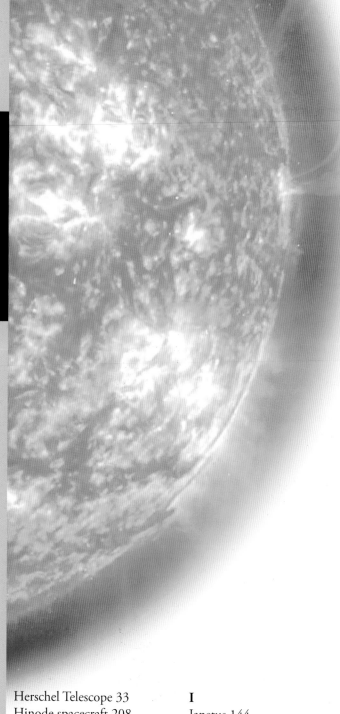

Acknowledgments

ACKNOWLEDGMENTS

The publisher would like to thank the following people for their help with making the book:
Ben Morgan, Kelsie Besaw, Mani Ramaswamy, Sai Prasanna, and Sukriti Kapoor for editorial assistance; Harish Aggarwal, Priyanka Sharma, and Saloni Singh for the jacket; Peter Radcliffe for design assistance; and Peter Bull for additional illustrations.

Smithsonian Enterprises:
Kealy Gordon, Product Development Manager
Jill Corcoran, Director, Licensed Publishing Sales
Brigid Ferraro, Vice President, Education and Consumer Products
Carol LeBlanc, President

Reviewers for the Smithsonian:
Ann Caspari, Education Specialist, Smithsonian's National Air and Space Museum
Rebecca Ljungren, Education Specialist, Smithsonian's National Air and Space Museum

Picture Credits
The publisher would like to thank the following for their kind permission to reproduce their photographs:

(Key: a-above; b-below/bottom; c-center; f-far; l-left; r-right; t-top)

1 Getty Images: Purestock. **2 Corbis:** Mark M. Lawrence (tr); Douglas Peebles (cra/Volcano). **Dorling Kindersley:** NASA (br). **NASA:** ESA (crb/Huygens); JPL (crb); JPL/ University of Arizona (cra). **Science Photo Library:** CCI Archives (cra/Herschel). **SOHO/EIT (ESA & NASA):** (cr). **3 Corbis:** Bettmann (ca/Chimps). **NASA:** (cb/Boot print) (cb/Discovery space shuttle) (br); A.Caulet St-ECF, ESA (cb); JPL-Caltech / SwRI / MSSS / Gerald Eichstadt / Sean Doran © CC NC SA (ca); ESA, and H. Richer (University of British Columbia) (tc); ESA, and the Hubble Heritage (STScI/AURA) -ESA/ Hubble Collaboration (cra); ESA, and The Hubble Heritage Team (STScI/AURA) (bc); GSFC (crb/Moon crater); MSFC (cr); Voyager 2 (crb) (cra/Antenna). **NRAO / AUI / NSF:** (c). **Reuters:** NASA (crb/Telescope). **SST, Royal Swedish Academy of Sciences, LMSAL:** (tr). **4 Corbis:** Bettmann (cra); NASA/ Science Faction (ca); NOAA (cla); Seth Resnick/ Science Faction (fcla). **SOHO/EIT (ESA & NASA):** (fcra). **4-5 Getty Images:** Stockbyte (tl). **5 Corbis:** Ed Darack/ Science Faction (fcla). **Getty Images:** Robert Gendler/Visuals Unlimited, Inc. (cla). **NASA:** MSFC (ca). **6-34 Chandra X-Ray Observatory:** X-ray: NASA/CXC/SAO; Optical: NASA/STScI; Infrared: NASA/JPL-Caltech/ Steward/O.Krause et al. (l). **6-7 Science Photo Library:** David Nunuk (Background). **7 Alamy Images:** Dennis Hallinan (fcla). **Chandra X-Ray Observatory:** X-ray: NASA/CXC/SAO; Optical: NASA/STScI; Infrared: NASA/JPL-Caltech/ Steward/O.Krause et al. (c). **Corbis:** Mark M. Lawrence (cl). **8 Alamy Images:** Dennis Hallinan (cl). **8-9 Alamy Images:** Dennis Hallinan (Background). **9 Corbis:** Mark M. Lawrence (l). **HubbleSite:** NASA / ESA / CXC / STScI / B. McNamara (University of Waterloo) (cr). **NASA:** (c); STS-51A (tr). **10 Getty Images:** (cl); Rob Atkins (tr). **NASA:** JPL-Caltech/R. Hurt (SSC) (cr). **10-11 NASA:** JPL-Caltech/C. Lonsdale (Caltech/IPAC) and the SWIRE Team (Background). **11 Science Photo Library:** Mark Garlick (c). **12-13 Science Photo Library:** Kaj R. Svensson. **14 Corbis:** Stapleton Collection (cl). **15 Corbis:** Paul Almasy (cl); Bettmann (tr) (r); Jose Fuste Raga (bc); Rob Matheson (t/Background); Seth Resnick/ Science Faction (c). **SOHO/EIT (ESA & NASA):** (tc). **16 Corbis:** Roger Ressmeyer (tr) (b). **16-17 Getty Images:** Stattmayer (t/Background). **17 Corbis:** Bettmann (cr) (clb); Roger Ressmeyer (cla); Jim Sugar (br). **18 Science Photo Library:** John Sanford. **19 Corbis:** Ed Darack/ Science Faction (tr); Roger Ressmeyer (crb). **European Southern Observatory (ESO) :** (bl). **Getty Images:** Joe McNally (clb). **Large Binocular Telescope Corporation :** (c). **Reuters:** NASA (cla). **TMT Observatory Corporation:** (br). **20 Corbis:** Matthias Kulka (ca); Mehau Kulyk/ Science Photo Library (bl); NASA/ JPL/ Science Faction (br). **21 Corbis:** Markus Altmann (fbl); NASA-CAL /Handout /Reuters (bl); NASA, ESA and The Hubble Heritage Team/ Handout/ Reuters (tc). **Science Photo Library:** David A. Hardy (c); NASA (br); NRAO / AUI / NSF (fbr); JPL/Caltech/Harvard-Smithsonian Center for Astrophysics (bc). **22 NASA:** JPL (bl). **22-23 NASA:** JPL-Caltech/ University of Arizona (c); JPL-Caltech/ IRAS / H. McCallon (b). **23 NASA:** JPL-Caltech/ K. Su (Univ. of Arizona) (tc). **Science Photo Library:** CCI Archives (tr); Robert Gendler (clb). **24 Courtesy of the NAIC - Arecibo Observatory, a facility of the NSF:** (cl). **24-25 NRAO / AUI / NSF:** (b). **25 NRAO / AUI / NSF:** (cla) (tr). **Science Photo Library:** Paul Wootton (tc). **26 (c) University Corporation for Atmospheric Research (UCAR) :** 2007 Copyright/ Carlye Calvin (cl). **ESA:** ECF (crb). **Max Planck Institute for Solar**

System Research: SUNRISE project/ P. Barthol (bl). **NASA:** Swift/ Stefan Immler, et al. (tr). **27 Chandra X-Ray Observatory:** Optical: Robert Gendler; X-ray: NASA/CXC/SAO/J.Drake et al. (clb). **ESA:** (ca). **NASA:** ESA (tl); SDO (tr); Courtesy of SOHO / MDI, SOHO / EIT & SOHO / LASCO consortia. SOHO is a project of international cooperation between ESA and NASA. (crb/Sun rays). **28 Getty Images:** NASA (l). **HubbleSite:** (br). **Science Photo Library:** Emilio Segre Visual Archives / American Institute Of Physics (cra). **29 Alamy Images:** Dennis Hallinan (b/ Earth). **Alamy Stock Photo:** UPI / NASA (crb). **Chris Hansen:** (br). **NASA:** (c); ESA and the Hubble SM4 ERO Team (tr). **NRAO / AUI / NSF:** (cb). **30 NASA:** STScI Digitized Sky Survey/Noel Carboni; NASA and The Hubble Heritage Team (STScI/ AURA) (bl); NASA, ESA, and J. Maíz Apellániz (Instituto de Astrofísica de Andalucía, Spain) (tr). **31 HubbleSite:** NASA, ESA and The Hubble Heritage Team (STScI/AURA) (cl). **NASA:** Courtesy NASA/JPL-Caltech (tl) (cr); JPL-Caltech/J. Bally (Univ. of Colo.) (br). **32 Chandra X-Ray Observatory:** NGST (bl). **ESA:** (bc); D. Ducros (tr). **32-33 Alamy Images:** Dennis Hallinan (Background). **33 Chandra X-Ray Observatory:** X-ray: NASA/ CXC/ SAO (cra); Optical: NASA/STScI; Infrared: NASA/JPL-Caltech/Steward/O.Krause et al. (fcra). **ESA:** D. Ducros, 2009 (bc). **HubbleSite:** NASA, ESA, and the Hubble Heritage Team (STScI/AURA) -ESA/ Hubble Collaboration (fcla). **NASA:** (bl) (br); JPL-Caltech (cla). **34 Global Oscillation Network Group (GONG) :** NSO/ AURA/ NSF/ MLSO/ HAO (cla). **Laser Interferometer Gravitational Wave Observatory (LIGO) :** (c). **National Science Foundation, USA:** Glenn Grant (br). **35 ALMA:** ESO/ NAOJ/ NRAO (crb) (clb). **NASA:** SOFIA (tl); Carla Thomas (cla). **The Sudbury Neutrino Observatory Institute (SNOI) :** Lawrence Berkeley National Laboratory for the SNO Collaboration (cr). **36-37 HubbleSite:** NASA, ESA, J. Hester and A. Loll (Arizona State University) (Background). **36-62 HubbleSite:** NASA, ESA, J. Hester and A. Loll (Arizona State University) (l). **37 HubbleSite:** (c); NASA, ESA, CXC, and JPL-Caltech (fcl). **NASA:** JPL-Caltech/R. Hurt (SSC) (cl). **38 Corbis:** Moodboard (bl). **38-39 HubbleSite:** NASA, ESA, and the Hubble Heritage Team (STScI/AURA) - ESA/Hubble Collaboration (c). **39 Alamy Images:** George Kelvin / PHOTOTAKE (cr) (crb) (crb). **Science Photo Library:** Detlev Van Ravensswaay (br). **40 Chandra X-Ray Observatory:** NASA/ CXC/ SAO/ P.Slane, et al. (bl). **43 © CERN :** Maximilien Brice (crb). **Corbis:** NASA/ epa (Background). **Getty Images:** Rob Atkins (fcra); Jeremy Horner (cra). **ESA:** Planck Collaboration (clb). **44-45 NASA:** ESA, H. Teplitz, and M. Rafelski (IPAC / Caltech), A. Koekemoer (STScI), R. Windhorst (Arizona State University), and Z. Levay (STScI) (Background). **45 Anglo Australian Observatory:** David Malin (br). **HubbleSite:** NASA, ESA, Y. Izotov (Main Astronomical Observatory, Kyiv, UA) and T. Thuan (University of Virginia) (crb). **NASA:** X-ray: CXC/Wesleyan Univ./R.Kilgard et al.; UV: JPL-Caltech; Optical: ESA/S. Beckwith & Hubble Heritage Team (STScI/AURA); IR: JPL-Caltech/ Univ. of AZ/R. Kennicutt (c). **Science Photo Library:** (c); JPL-Caltech/CTIO

(bc). **46 NASA:** JPL-Caltech (bl) (br). **Science Photo Library:** Volker Springel / Max Planck Institute For Astrophysics (cl). **46-47 NASA:** JPL-Caltech/STScI/CXC/UofA/ ESA/AURA/JHU (c). **47 European Southern Observatory (ESO) :** (bl). **NASA:** Al Kelly (JSCAS/NASA) & Arne Henden (Flagstaff/USNO) (bc); ESA, A. Aloisi (STScI / ESA), Hubble Heritage (STScI / AURA) - ESA/Hubble Collaboration (fbl); The Hubble Heritage Team (STScI/AURA) (ca); ESA, and R.A. Lucas (STScI) (cra). **48-49 HubbleSite:** NASA and The Hubble Heritage Team (STScI/AURA, x). **50 European Southern Observatory (ESO) :** Yuri Beletsky (cl). **Science Photo Library:** Chris Butler (bl). **50-51 NASA:** JPL-Caltech/R. Hurt (SSC) (c); CXC/ MIT/ Frederick K. Baganoff et al. (crb). **51 NASA:** CXC/ UMass/ D. Wang et al. (tr); JPL-Caltech/ R. Hurt (SSC) (bc); JPL-Caltech/ S. V. Ramirez (NExScI/ Caltech) , D. An (IPAC/ Caltech) , K. Sellgren (OSU) (clb); NASA/ CXC/ M.Weiss (cra). **52 Chandra X-Ray Observatory:** NASA/ SAO/ CXC (crb). **NASA:** JPL-Caltech /M. Meixner (STScI) & the SAGE Legacy Team (cl). **53 CSIRO:** Dallas Parr (bl). **ESA:** Hubble and Digitized Sky Survey 2 (tl); NASA, ESO and Danny LaCrue (cra). **NASA:** ESA, and the Hubble Heritage Team (STScI/AURA) (tr). **54 Science Photo Library:** Mark Garlick (br); MPIA-HD, BIRKLE, SLAWIK (c). **55 NASA:** Adam Block/ NOAO/ AURA/ NSF (c); JPL-Caltech/D. Block (Anglo American Cosmic Dust Lab, SA) (tr); JPL-Caltech/Univ. of Ariz. (cl); Paul Mortfield, Stefano Cancelli (br); UMass/Z. Li & Q.D.Wang (tc). **56-57 NASA:** JPL-Caltech/ ESA/ CXC/ STScI. **58 NASA:** X-ray: NASA / CXC/ CfA/E. O'Sullivan Optical: Canada-France-Hawaii-Telescope/ Coelum (c). **58-59 Courtesy of Dr Stelios Kazantzidis (Center for Cosmology and Astro-Particle Physics, The Ohio State University) :** (b/Spiral galaxy collision); NASA, ESA, and the Hubble Heritage Team (STScI/ AURA) (tr); NASA, ESA, and the Hubble Heritage Team (STScI/ AURA) -ESA/ Hubble Collaboration (crb); NASA, ESA, Richard Ellis (Caltech) and Jean-Paul Kneib (Observatoire Midi-Pyrenees, France) (clb); NASA, H. Ford (JHU) , G. Illingworth (UCSC/LO) , M.Clampin (STScI) , G. Hartig (STScI) , the ACS Science Team, and ESA (cr). **59 HubbleSite:** NASA, ESA, CXC, C. Ma, H. Ebeling, and E. Barrett (University of Hawaii/ IfA) , et al., and STScI (tl). **60 Corbis:** STScI/ NASA (crb). **Till Credner , Allthesky. com:** (Background). **HubbleSite:** (bl). **Science Photo Library:** NRAO / AUI / NSF (cr). **61 Chandra X-Ray Observatory:** X-ray: NASA/ CXC/Univ. of Maryland/A.S. Wilson et al.; Optical: Pal.Obs. DSS; IR: NASA/JPL-Caltech; VLA: NRAO/AUI/NSF (bl). **HubbleSite:** John Hutchings (Dominion Astrophysical Observatory) , Bruce Woodgate (GSFC/NASA), Mary Beth Kaiser (Johns Hopkins University),

Steven Kraemer (Catholic University of America), the STIS Team., and NASA (tl). **NRAO / AUI / NSF:** Image courtesy of National Radio Astronomy Observatory / Associated Universities, Inc. / National Science Foundation (cra). **Science Photo Library:** NASA / ESA / STSCI / J.BAHCALL, PRINCETON IAS (crb). **62 Science Photo Library:** Mike Agliolo (crb); Volker Springel / Max Planck Institute For Astrophysics (cl). **62-63 Science Photo Library:** Lynette Cook. **63 NASA:** X-ray: NASA / CXC / Caltech / A. Newman et al. / Tel Aviv / A. Morandi & M. Limousin; Optical: NASA / STScI, ESO / VLT, SDSS (bc). **Science Photo Library:** M. Markevitch/ CXC/ CFA/ NASA (bl). **64-65 Getty Images:** AFP/ Jim Watson (Background). **64-88 Dorling Kindersley:** ESA - ESTEC (l). **65 Corbis:** Bettmann (fcl). **ESA:** (c). **US Geological Survey:** Astrogeology Team (cl). **66 Getty Images:** Sir Godfrey Kneller (c). **NASA:** KSC (l); United Launch Alliance/ Pat Corkery (r). **67 NASA:** Bill Ingalls (c); Pratt & Whitney Rocketdyne (tr). **68-69 John Kraus. 70 Alamy Images:** Linda Sikes (br). **Corbis:** NASA/CNP (c). **Science Photo Library:** Mark Garlick (cl). **71 Alamy Images:** Stock Connection Blue (c). **Corbis:** (cr); Bettmann (tl). **Getty Images:** NASA (clb). **Science Photo Library:** NASA (crb). **72 NASA:** (tr); KSC (crb); MSFC / KSC (cl). **73 ESA:** (cl). **EUROCKOT Launch Services GmbH:** (cra). **Getty Images:** Space Imaging (bl). **NASA:** Victor Zelentsov (tl). **Courtesy Sea Launch:** (br). **74 ESA:** CNES/ Arianespace/ Photo optique video du CSG (clb) (cb); Service Optique CSG (cr). **74-75 ESA:** CNES/ Arianespace/ Photo optique video du CSG (t). **75 ESA:** CNES/ Arianespace/ Photo optique video du CSG (bc) (cr); Service Optique CSG (cl). **76 Corbis:** Alain Nogues/ Sygma (br). **NASA:** JPL (clb). **77 NASA:** (b). **78 Corbis:** Bettmann (b). **ESA:** D. Ducros (c). **NASA:** Goddard Space Flight Center/ MODIS Rapid Response Team/ Jeff Schmaltz (cr). **79 CNES:** Illustration P.Carril - Mars 2003 (clb). **© EADS :** Astrium (crb). **ESA:** J. Huart (cra). **80-81 USGS:** Courtesy of the U.S. Geological Survey. **82 Getty Images:** Ludek Pesek (b). **NASA:** NSSDC (tl). **Science Photo Library:** Detlev Van Ravenswaay (c). **83 NASA:** Ames Research Center (cra); JPL (tl); NSSDC (clb). **Science Photo Library:** NASA / JPL (bc). **US Geological Survey:** Astrogeology Team (fclb). **Wikimedia Commons:** Daderot (br). **84 ESA:** (c). **NASA:** (crb). **85 CNES:** Illustration D. Ducros - 1998 (cr). **ESA:** (tr). **NASA:** (cb) (bc) (br). **86 NASA:** ISRO/ JPL-Caltech/ USGS/ Brown Univ. (bl). **Science Photo Library:** Indian Space Research Organisation (r). **87 CBERS:** INPE (cra). **Corbis:** Li Gang/ Xinhua Press (tc). **Alamy Stock Photo:** Xinhua (cl). **Akihoro Ikeshita:** (clb). **Courtesy of JAXA:** NHK (b/ Background) (cb). **88 ESA:** AOES Medialab/ ESA 2002 (cb). **Science Photo Library:** David A. Hardy, Futures: 50 Years In Space (ca). **89 Courtesy of JAXA:** (crb). **Science Photo Library:** David A. Hardy (clb); NASA (tl) (cra). **90-114 Dorling Kindersley:** NASA (l). **90-91 Getty Images:** NASA/ National Geographic (Background). **91 Corbis:** Bettmann (cl). **NASA:** (fcl). **SpaceX:** (c). **92 Corbis:** Bettmann (cl) (cr); NASA - digital version copyright/Science Faction (bl). **NASA:** 5909731 / MSFC-5909731 (cra). **92-93 Corbis:** Bettmann (Background). **93 Corbis:** Bettmann (cla) (bc) (cr); Karl Weatherly (cb). **Dorling Kindersley:** Bob Gathany (tl). **NASA:** (clb); MSFC (tr). **94 NASA:** ESA (r); Robert Markowitz/ Mark Sowa (bc). **95 ESA:** (cla) (cra); ASI-Star City (crb). **NASA:** (cl) (br) (clb); Bill Ingalls (tl). **Science Photo Library:** NASA (cr). **96 NASA:** JSC (clb) (b) (tr). **97 Dorling Kindersley:** NASA (clb). **NASA:** JSC (crb) (br) (fbr). **Science Photo Library:** NASA (t). **98 NASA:** (cl) (bc). **Science Photo Library:** NASA (crb) (br). **99 NASA:** (tl) (bc) (tr). **Wikimedia Commons:** Aliazimi (bl). **100 Alamy Images:** RIA Novosti (cl). **Corbis:** Bettmann (c); Hulton-Deutsch Collection (tr). **Getty Images:** Hulton Archive (ca). **NASA:** 5909731 / MSFC-5909731 (br). **101 Corbis:** Roger Ressmeyer (tr). **NASA:** Kennedy Space Center (bl). **Dreamstime.com:** roblan (fbr). **Science Photo Library:** Power And Syred (crb). **102-103 NASA:** (Background). **104 Alamy Images:** RIA Novosti (cla). **The Kobal Collection:** MGM (crb). **NASA:** (clb). **104-105 Science Photo Library:** NASA (b). **105 NASA:** (tl) (cra). **106 NASA:** (cl) (br). **107 NASA:** (br) (cla) (cr) (tr). **108 NASA:** (cra) (cb) (crb). **109 Corbis:** Bettmann (tc). **NASA:** (c); MSFC (bl) (clb). **Science Photo Library:** NASA (cra). **110 Alamy Stock Photo:** Blue Origin (bl). **NASA:** KSC (br). **NASA:** Scaled Composites (ca). **110-111 Corbis:** Ed Darack/ Science Faction (Background). **111 Bigelow Aerospace :** (bl). **Getty Images:** Daniel Berehulak (ca). **courtesy Virgin Galactic:** (tl) (cra). **SpaceX:** (crb). **112 Reaction Engines Limited / Adrian Mann:** Reaction Engines Ltd develops SKYLON, a space plane which evolved from the HOTOL project (b). **Science Photo Library:** Richard Bizley (cra). **113 Agence France Presse:** (crb). **iStockphoto.com: 3DSculptor** (cl). **NASA:** DFRC/ Illustration by Steve Lighthill (b). **SpaceX:** (tr). **114 Alamy Images:** Pat Eyre (crb). **Corbis:** James Marshall (cb). **ESA:** S. Corvaja (bl). **Science Photo Library:** Sinclair Stammers (cra). **115 Alamy Images:** Photos 12 (cr). **NASA:** MSFC (clb). **PA Photos:** AP/ NASA (br). **Science Photo Library:** Victor Habbick Visions (t). **116-117 NASA:** JPL/ University of Arizona (Background). **116-162 Dorling Kindersley:** NASA /Finley Holiday Films (l). **117 Corbis:** Dennis di Cicco (c). **NASA:** JPL-Caltech / SwRI / MSSS / Gerald Eichstadt / Sean Doran © CC NC SA (cl). **118 NASA:** JPL-Caltech / T. Pyle (SSC) (c). **Science Photo Library:** Detlev Van Ravenswaay (crb). **119 David A. Hardy :** PPARC (br). **Julian Baum:** (clb). **120 HubbleSite:** Reta Beebe (New Mexico State University) / NASA (cb); NASA, ESA, L. Sromovsky and P. Fry (University of Wisconsin) , H. Hammel (Space Science Institute) , and K. Rages (SETI Institute) (crb). **NASA:** (clb/Earth); NASA and The Hubble Heritage Team (STScI / AURA) Acknowledgment: R.G. French (Wellesley College), J. Cuzzi (NASA / Ames), L. Dones (SwRI), and J. Lissauer (NASA / Ames) (crb/ Saturn). **120-121 NASA:** JPL-Caltech (solar system planets). **121 NASA:** Johns Hopkins University Applied Physics Laboratory / Southwest Research Institute (bc). **122 NASA:** JPL / USGS (r); NASA and The Hubble Heritage Team (STScI / AURA) Acknowledgment: R.G. French (Wellesley College), J. Cuzzi (NASA / Ames), L. Dones (SwRI), and J. Lissauer (NASA / Ames) (crb/ Saturn). **123 Getty Images:** Dieter Spannknebel (tl). **NASA:** Johns Hopkins University Applied Physics Laboratory / Carnegie Institution of Washington (clb). NSSDC/ GSFC (ca). **Science Photo Library:** M. Ledlow Et Al / NRAO / AUI / NSF (cb). **SOHO/EIT (ESA & NASA) :** (cr). **124 NASA:** NASA and The Hubble Heritage Team (STScI / AURA) Acknowledgment: R. G. French (Wellesley College), J. Cuzzi (NASA / Ames), L. Dones (SwRI), and J. Lissauer (NASA / Ames) (clb/ Saturn). **124-125 Science Photo Library:** NASA (tc). **125 ESA:** MPS/ Katlenburg-Lindau (crb). **NASA:** JPL (cla); NSSDC (bl) (bc). **126 NASA:** JPL (cra) (b) (clb). **127 ESA:** (crb). **NASA:** Ames Research Center (tr); JPL (tl); JPL-Caltech (cra) (c) (cl). **Science Photo Library:** David P. Anderson, SMU/ Nasa (cb). **128 ESA:** DLR/ FU Berlin (G. Neukum) (bc). **NASA:** (cra); ESA (cr); JPL (cb); JPL/ Malin Space Science Systems (br); NSSDC (bl). **129 Getty Images:** Time & Life Pictures (clb). **NASA:** GSFC (r); JPL /MSSS (tl); JPL/ Malin Space Science Systems (ca); NASA and The Hubble Heritage Team (STScI / AURA) Acknowledgment: R. G. French (Wellesley College), J. Cuzzi (NASA / Ames), L. Dones (SwRI), and J. Lissauer (NASA / Ames) (bl/ Saturn). **130 Corbis:** Lowell Georgia (bc); JPL / USGS (r); JPL /MSSS (cb). **NASA:** JPL/ University of Arizona (cl); JPL/ Cornell (fbr). **Science Photo Library:** NASA (br). **131 ESA:** G. Neukum (FU Berlin) et al./ Mars Express/ DLR (cra); JPL-Caltech (cb/Rover). **NASA:** JPL-Caltech (clb); JPL-Caltech/ MSSS (crb, br); JPL/ Cornell (t) (bc) (ca). **132-133 NASA:** HiRISE/ JPL/ University of Arizona. **134 Alamy Images:** Mary Evans Picture Library (tr). **Science Photo Library:** Chris Butler (bc). **NASA:** JPL-Caltech (crb). **135 Rex by Shutterstock:** Uncredited / AP (tc). **Science Photo Library:** Chris Butler (tr); Henning Dalhoff / Bonnier Publications (crb); D. Van Ravenswaay (cl). **136 HubbleSite:** NASA/ESA, John Clarke (University of Michigan) (tr). **NASA:** JPL-Caltech / SwRI / MSSS / Gerald Eichstadt / Sean Doran © CC NC SA (br). **137 Corbis:** NASA-JPL-Caltech - digital versi/Science Faction (r). **NASA:** JPL-Caltech / SwRI / MSSS / Gabriel Fiset (clb); JPL/ Cornell University (cla); NASA and The Hubble Heritage Team (STScI / AURA) Acknowledgment: R. G. French (Wellesley College), J. Cuzzi (NASA / Ames), L. Dones (SwRI), and J. Lissauer (NASA / Ames) (bl/ Saturn). **138 Corbis:** Bettmann (tr); JPL / USGS (b). **NASA:** JPL/ University of Arizona (c); JPL/ Brown University (bl); JPL/ DLR (tr); JPL/ University of Arizona (cr). **139 NASA:** JPL (cla) (bc) (cl) (crb). **140 NASA:** JPL-Caltech (crb); MSFC (bl); NASA and The Hubble Heritage Team (STScI / AURA) Acknowledgment: R.G. French (Wellesley College), J. Cuzzi (NASA / Ames), L. Dones (SwRI), and J. Lissauer (NASA / Ames) (cb/ Saturn). **141 NASA:** JPL (clb); JPL-Caltech (cr) (cl) (br); JPL/ Space Science Institute (clb). **142 NASA:** JPL/ STScI (tr). **Science Photo Library:** D. Van Ravenswaay (crb); NASA, ESA, J. Clarke (Boston University), and Z. Levay (STScI) (c). **142-143 NASA:** NASA and The Hubble Heritage Team (STScI / AURA) Acknowledgment: R. G. French (Wellesley College), J. Cuzzi (NASA / Ames), L. Dones (SwRI), and J. Lissauer (NASA / Ames) (b). **143 Corbis:** NASA - digital version copyright/ Science Faction (tc); STScI/ NASA (bc). **Science Photo Library:** NASA/ JPL/ University Of Arizona (tl). **NASA:** NASA and The Hubble Heritage Team (STScI / AURA) Acknowledgment: R. G. French (Wellesley College), J. Cuzzi (NASA / Ames), L. Dones (SwRI), and J. Lissauer (NASA / Ames) (bl/ Saturn). **144 Alamy Images:** The Print Collector (tr); JPL / USGS (bc). **NASA:** JPL/ Space Science Institute (clb) (cl) (crb); NASA and The Hubble Heritage Team (STScI / AURA) Acknowledgment: R.G. French (Wellesley College), J. Cuzzi (NASA / Ames), L. Dones (SwRI), and J. Lissauer (NASA / Ames) (cb/ Saturn). **NRAO / AUI / NSF:** (cr). **144-145 NASA:** JPL/ Space Science Institute (tc). **145 ESA:** (br); NASA/ JPL/ University of Arizona (tr) (cb) (crb). **NASA:** JPL-Caltech (cla); JPL (bl); JPL/ GSFC/ Space Science Institute (clb); JPL/ University of Arizona (ca). **146-147 NASA:** JPL/ Space Science Institute. **148 Getty Images:** John Russell (cl). **W.M. Keck Observatory:** Lawrence Sromovsky, (Univ. Wisconsin-Madison) (cb). **NASA:** JPL (br); NSSDC (l). **149 NASA:** GSFC (br); JPL (c); JPL / USGS (bl); JPL-Caltech (cla) (cr) (fcr); NSSDC (cra); NASA and The Hubble Heritage Team (STScI / AURA) Acknowledgment: R. G. French (Wellesley College), J. Cuzzi (NASA / Ames), L. Dones (SwRI), and J. Lissauer (NASA / Ames) (tr/ Saturn). **150 NASA:** (bl); Voyager 2 (c). **151 NASA:** (cra); JPL (bl); JPL / USGS (cla); NASA and The Hubble Heritage Team (STScI / AURA) Acknowledgment: R. G. French (Wellesley College), J. Cuzzi (NASA / Ames), L. Dones (SwRI), and J. Lissauer (NASA / Ames) (br/ Saturn). **Science Photo Library:** Royal Astronomical Society (cr). **152 NASA:** Johns Hopkins University Applied Physics Laboratory / Southwest Research Institute / Roman Tkachenko (clb). **152-153 NASA:** JHUAPL / SwRI (tc). ESA and G. Bacon (STScI) (b). **153 HubbleSite:** ESA, H. Weaver (JHU/ APL) , A. Stern (SwRI) , and the HST Pluto Companion Search Team (clb). **Getty Images:** Universal Images Group (cr). **NASA:** JHUAPL / SwRI (cl). **154 Corbis:** Dennis di Cicco (b). **155 Corbis:** Jonathan Blair (bl); Gianni Dagli Orti (cla). **HubbleSite:** NASA / ESA / M. Wong (Space Telescope Science Institute, Baltimore, Md.) / H. B. Hammel (Space Science Institute, Boulder, Colo.) / Jupiter Impact Team (cr). **Science Photo Library:** Mark Garlick (br); Gordon Garradd (tr); NASA / ESA / STSCI / H. Weaver & T. Smith (c). **156 Corbis:** NASA (crb); Roger Ressmeyer (c). **Dorling Kindersley:** ESA (cra). **ESA:** SOHO (clb). **NASA:** JPL (br). **157 HubbleSite:** NASA, ESA, P. Feldman (Johns Hopkins University) and H. Weaver (Johns Hopkins University Applied Physics Laboratory) (tr); JPL/ UMD (cra). **NASA:** JPL (tl); MSFC (cl). **Science Photo Library:** Erik Viktor (crb). **ESA:** Rosetta / NAVCAM, CC BY SA 3.0 IGO (bl). **158 ICSTARS Astronomy:** Vic & Jen Winter. **159 Corbis:** Tony Hallas/ Science Faction (br). **Spacewatch:** Jim Scotti, Spacewatch (cl). **Alex Alishevskikh:** cyberborean (tr). **Kwon, O Chul:** (cr). **160 Corbis:** Hans Schmied (ca). **Science Photo Library:** Mark Garlick (cla). **160-161 Corbis:** Bryan Allen (b). **161 Dorling Kindersley:** The Natural History Museum, London (ca). **Galaxy Picture Library:** UWO/ University of Calgary/ Galax (tr). **NASA:** Ted Bunch/ JPL (fcra); M. Elhassan/ M. H. Shaddad/ P. Jenniskens (crb); Michael Farmer/ JPL (cr); JPL-Caltech / Cornell University (cl). **162 Selden E. Ball:** Cornell University (cl). **NASA:** JPL-Caltech / MSSS / JHU-APL / Brown Univ. (cr). **Science Photo Library:** Christian Darkin (b); NASA (ca); T. Stevens & P. Mckinley, Pacific Northwest Laboratory (cra). **163 NASA:** (tc); JPL / USGS (clb); JPL/ University of Arizona (cl); JPL/ University of Arizona/ University of Colorado (tr); NOAA (cr). **William Falconer-Beach:** (clb). **Science Photo Library:** Mark Garlick (br); US Geological Survey (crb). **164-165 Science Photo Library:** Planet Observer (Background). **164-176 Dorling Kindersley:** NASA (l). **165 Corbis:** Momatiuk - Eastcott (fcl); Douglas Peebles (c). **Getty Images:** Barcroft Media (cl). **166 Dorling Kindersley:** Planetary Visions Ltd (clb). **NASA:** NASA and The Hubble Heritage Team (STScI / AURA) Acknowledgment: R. G. French (Wellesley College), J. Cuzzi (NASA / Ames), L. Dones (SwRI), and J. Lissauer (NASA / Ames) (cl/ Saturn) **166-167 NASA:** (c). **167 NASA:** (bc) (tr/Earth); MSFC (cr). **168-169 Alamy Images:** Rolf Nussbaumer Photography. **170 Alamy Images:** Alaska Stock LLC (b). **NASA:** JPL (cl). **171 iStockphoto.com:** Janrysavy (cl) (cb) (cr) (fcrb). **NASA:** GSFC (bl); MODIS Ocean Science Team (br). **Science Photo Library:** European Space Agency (c). **172 Corbis:** Douglas

ACKNOWLEDGMENTS

Peebles (crb). **172-173 Corbis:** Galen Rowell (b). **173 Corbis:** Momatiuk - Eastcott (ca). **Science Photo Library:** Bernhard Edmaier (crb); David Parker (br); Ron Sanford (tr). **174 Corbis:** Bryan Allen (c); Hinrich Baesemann / DPA (cl). **NASA:** (tr). **Science Photo Library:** Detlev Van Ravensway (br). **175 Corbis:** (br); Mike Hollingshead / Science Faction (bl); Gerolf Kalt (clb); NOAA (cr). **Science Photo Library:** David R. Frazier (tl). **176 Dorling Kindersley:** The Royal Museum of Scotland, Edinburgh (br). **Science Photo Library:** Lynette Cook (Volcanoes); Henning Dalhoff / Bonnier Publications (clb). **177 Alamy Images:** Amberstock (tl). **Dorling Kindersley:** Jon Hughes (bl) (bc). **ESA:** (crb). imagequestmarine. com: Peter Batson (cb). **NOAA:** Office of Ocean Exploration; Dr. Bob Embley, NOAA PMEL, Chief Scientist (cl). **Science Photo Library:** Victor Habbick Visions (tr); P. Rona/ OAR/ National Undersea Research Program/ NOAA (cl). **178-179 Alamy Images:** Melba Photo Agency (Background). **179 NASA:** (cl) (c). **180 Alamy Images:** Patrick Eden (b). **Science Photo Library:** Andrew J. Martinez (cra) (fcra). **181 Corbis:** William Radcliffe/ Science Faction (cr). **Science Photo Library:** Planetary Visions Ltd (bc). **NASA:** NASA and The Hubble Heritage Team (STScI / AURA) Acknowledgment: R. G. French (Wellesley College), J. Cuzzi (NASA / Ames), L. Dones (SwRI), and J. Lissauer (NASA / Ames) (tr/ Saturn). **NASA:** VGL/ amanaimagesRF (fcra). **NASA:** Image courtesy of the Image Science & Analysis Laboratory, NASA Johnson Space Center (cra). **Science Photo Library:** NOAO (tc); David Nunuk (crb). **182-192 Dorling Kindersley:** NASA (l). **183 Corbis:** Tom Fox/ Dallas Morning News (crb); Reuters (tl). **Getty Images:** VGL/ amanaimagesRF (cr). **NASA:** Image courtesy of the Image Science & Analysis Laboratory, NASA Johnson Space Center (fcr). **Science Photo Library:** NOAO (c). **184 NASA:** JSC (cl). **184-185 Getty Images:** Stocktrek RF (c). **Moonpans.com:** (b). **185 Getty Images:** SSPL (crb). **NASA:** JSC (cb); MSFC (cla); NSSDC (cra). **186 Getty Images:** Viewstock (bl). **187 NASA:** (cr) (cb) (crb); Neil A. Armstrong (cl) (br); **Science Photo Library:** D. Van Ravensway (clb); Ria Novosti (bl). **188 Moonpans.com:** Charlie Duke (b). **NASA:** JSC (tl) (cl). **189 Corbis:** (tl); Roger Ressmeyer (bl). **NASA:** (crb); Charlie Duke (cra); JSC (ca). **Science Photo Library:** NASA (tl). **190 Corbis:** NASA / Roger Ressmeyer (bl). **NASA:** MSFC (cla) (clb). **190-191 NASA. 192 Courtesy of JAXA:** (cra). **NASA:** (cb) (cr); JPL-Caltech / MIT / GSFC (clb); JPL (br); NSSDC (l). **193 Courtesy of JAXA:** (tl). **Shutterstock:** testing (clb). **NASA:** (b). **Alamy Stock Photo:** Xinhua (cra). **194-195 NASA:** SOHO. **194-208 Alamy Images:** Brand X Pictures (l). **195 NASA:** GSFC / TRACE (cl); TRACE (c). **SST, Royal**

Swedish Academy of Sciences, LMSAL: (fcl). **196 NASA:** (cl); SOHO (fbr); Goddard Space Flight Center Conceptual Image Lab (tl); NASA and The Hubble Heritage Team (STScI / AURA) Acknowledgment: R. G. French (Wellesley College), J. Cuzzi (NASA / Ames), L. Dones (SwRI), and J. Lissauer (NASA / Ames) (crb/ Saturn). **Science Photo Library:** John Chumack (cr). **SOHO/EIT (ESA & NASA) :** (br). **197 NASA:** SOHO. **198 (c) University Corporation for Atmospheric Research (UCAR) :** Illustration by Mark Miesch (tr). **NASA:** (b). **199 NASA:** (cl); GSFC / A. Title (Stanford Lockheed Institute) / TRACE (tl); GSFC / SOHO (br). **200 NASA:** TRACE (bl). **200-201 NASA:** Steve Albers / Dennis di Cicco / Gary Emerson. **201 NASA:** (br); JPL-Caltech (tl); SOHO (cra). **202 NASA:** GSFC (b). **SST, Royal Swedish Academy of Sciences, LMSAL:** (tr). **203 NASA:** GSFC (clb); SOHO / ESA (t); SOHO / MSFC (c) (br). **204-205 Getty Images:** Moment / Steffen Schnur. **206-207 SOHO (ESA & NASA):** (t). **207 David Hathaway/Stanford University:** (cb). **208 Corbis:** Bettmann (tr). **Science Photo Library:** Royal Astronomical Society (cr). **NASA:** MSFC (crb). **Alamy Images:** BWAC Images (bl). **208-209 National Science Foundation's National Solar Observatory:** NSO / AURA / NSF (b). **209 Reuters:** Ho New (tr). **210-211 HubbleSite:** NASA / ESA / A. Nota (STScI / ESA). **210-238 HubbleSite:** NASA, ESA, and Martino Romaniello (European Southern Observatory, Germany) (l). **211 Corbis:** Stapleton Collection (cl); NASA / ESA / HEIC / The Hubble Heritage Team / STScI / AURA (c). **HubbleSite:** NASA / ESA / J. Hester (ASU) (fcl). **212 HubbleSite:** NASA / ESA / M. Robberto (Space Telescope Science Institute / ESA) / Hubble Space Telescope Orion Treasury Project Team (bc). **213 Anglo Australian Observatory:** D. Malin (AAO) / AATB / UKS Telescope (ftr). **NASA:** (tr); Compton Gamma Ray Observatory / GSFC (tc); ESA / H. Bond (STScI) / M. Barstow (University of Leicester) (ftl). **Science Photo Library:** European Space Agency (cla) (bl) (br) (clb) (crb) (r); NASA / A. Caulet / St-ECF / ESA (tr). **214 HubbleSite:** NASA / Jeff Hester (Arizona State University) (tl). **214-215 HubbleSite:** heic0506b / opo0512b. **215 HubbleSite:** A. Caulet (ST-ECF, ESA) / NASA (cla); NASA / ESA / SM4 ERO Team (br). **NASA:** ESA (tr); Ryan Steinberg & Family / Adam Block / NOAO / AURA / NSF (tl). **216 HubbleSite:** NASA / ESA / H. E. Bond (STScI) / The Hubble Heritage Team (STScI / AURA). **216-217 HubbleSite:** NASA / ESA / the Hubble Heritage Team (STScI / AURA). **218 Anglo Australian Observatory:** David Malin (tl); NASA / ESA / Hans Van Winckel (Catholic University of Leuven, Belgium) / Martin Cohen (University of California, Berkeley) (br); NASA / ESA / HEIC / The Hubble Heritage Team / STScI / AURA (bl);

NASA / Jon Morse (University of Colorado) (cr). **HubbleSite:** NASA / ESA / Andrea Dupree (Harvard-Smithsonian CfA) / Ronald Gilliland (STScI) (ca). **219 Chandra X-Ray Observatory:** X-ray: NASA / CXC / Rutgers / G.Cassam-Chenaï / J.Hughes et al.; Radio: NRAO / AUI / NSF / GBT / VLA / Dyer, Maddalena & Cornwell / Optical: Middlebury College / F. Winkler / NOAO / AURA / NSF / CTIO Schmidt & DSS (cr); NASA (bc); NASA / Andrew Fruchter / ERO Team - Sylvia Baggett (STScI) / Richard Hook (ST-ECF) / Zoltan Levay (STScI) (br). **HubbleSite:** NASA / The Hubble Heritage Team (STScI / AURA) / W. Sparks (STScI) / R. Sahai (JPL) (bl). **220 HubbleSite:** NASA / ESA / The Hubble Heritage Team (STScI / AURA) / P. McCullough (STScI). **NASA:** NOAO / T. A. Rector / U. Alaska / T. Abbott / AURA / NSF (br). **Naval Research Lab.:** Rhonda Stroud / Nittler (2003) (cra). **221 HubbleSite:** NASA / K.L. Luhman (Harvard-Smithsonian Center for Astrophysics, Cambridge, Mass.) / G. Schneider, E. Young, G. Rieke, A. Cotera, H. Chen, M. Rieke, R. Thompson (Steward Observatory, University of Arizona, Tucson, Ariz.) (bl). **NASA:** NOAO / T. A. Rector / U. Alaska / WIYN / AURA / NSF / GSFC (b). **222 HubbleSite:** NASA / ESA / G. Bacon (STScI) (bc). **NASA:** CXC / SAO / M. Karovska et al; (cl). **223 HubbleSite:** NASA / ESA / (STScI / AURA) / J. Maíz Apellániz (Institute of Astrophysics of Andalucía, Spain). **224-225 HubbleSite:** NASA / ESA / the Hubble Heritage Team (STScI / AURA) / A. Cool (San Francisco State University) / J. Anderson (STScI). **225 HubbleSite:** NASA / ESA / H. Richer (University of British Columbia) (bl). **NASA:** GSFC (crb). **226 ESA:** NASA / L. Ricci (ESO) (ca) (br) (cr) (fbl) (fclb). **HubbleSite:** NASA / ESA / M. Robberto (Space Telescope Science Institute / ESA) / the Hubble Space Telescope Orion Treasury Project Team (crb). **NASA:** JPL-Caltech (fcra). **227 NASA:** JPL (br). **ESA:** A. M. Lagrange et al. (cra); M. Kornmesser (clb). **228 HubbleSite:** NASA / H. Richer (University of British Columbia). **NASA:** (cl); CXC / M. Weiss (bl); JPL-Caltech (crb). **229 NASA:** (tr); Dana Berry (br); CXC / SAO / F. Seward (c); JPL (clb). **230 Dorling Kindersley:** NASA (bc) (crb) (fcr). **HubbleSite:** ESA, NASA, and Felix Mirabel (French Atomic Energy Commission and Institute for Astronomy and Space Physics / Conicet of Argentina) (cl). **EHT Collaboration:** (tr). **Science Photo Library:** CXC / AlfA / D. Hudson and T. Reiprich et al.; NRAO / VLA / NRL / NASA (bl). **231 Science Photo Library:** European Space Agency. **232 Science Photo Library:** David Nunuk (b). **232-233 Science Photo Library:** Larry Landolfi. **233 Alamy Images:** Tony Craddock / Images Etc Ltd (ca). **Corbis:** Jay Pasachoff / Science Faction (bl). **234 Corbis:** Stapleton Collection (cl) (tr). **Getty Images:** The Bridgeman Art Library / Andreas

Cellarius (br). **235 Science Photo Library:** Pekka Parviainen (tl). **236 Science Photo Library:** Davide De Martin (tr); NASA / JPL-Caltech / STSCI (cl); Eckhard Slawik (cr). **238 Corbis:** Radius Images (cl). **Getty Images:** Robert Gendler/Visuals Unlimited, Inc. (tr); Stone / World Perspectives (cr). **240 Corbis:** Bettmann (cla) (fcrb); Gianni Dagli Orti (clb); Stapleton Collection (fclb). **Getty Images:** Science & Society Picture Library (cr). **Dorling Kindersley:** NASA /Finley Holiday Films (fcr); Rough Guides (fcla). **Science & Society Picture Library:** (cra). **Science Photo Library:** Chris Butler (crb). **NASA:** NASA and The Hubble Heritage Team (STScI / AURA) Acknowledgment: R. G. French (Wellesley College), J. Cuzzi (NASA / Ames), L. Dones (SwRI), and J. Lissauer (NASA / Ames) (bc). **240-241 iStockphoto.com:** Gaffera. **241 Corbis:** Bettmann (cra); NASA - digital version copyright/Science Faction (cl). **Dorling Kindersley:** Anglo-Australian Observatory/ David Malin (clb). **Getty Images:** Time & Life Pictures (cb). **NASA:** ESA and G. Bacon (STScI) (tl). **Science Photo Library:** NASA / JPL (br). **242 Alamy Images:** Stock Connection Blue / Novastock (cra). **Dorling Kindersley:** The Science Museum, London (ca). **NASA:** (fcla); JPL (bc); JPL-Caltech (cb). **Science Photo Library:** Ria Novosti (bl) (ftl); Detlev Van Ravensway (fcrb). **242-243 iStockphoto.com:** Gaffera. **243 Corbis:** Reuters (clb); JPL / Scaled Composites (tr). **NASA:** JPL (crb) (tl); NASA / ESA / STSCI / H. Ford Et Al (fcla). **Science Photo Library:** NASA (fcr); Friedrich Saurer (br); Detlev Van Ravensway (ca). **244 Science Photo Library:** Henning Dalhoff / Bonnier Publications. **245 HubbleSite:** NASA, ESA, and The Hubble Heritage Team (STScI / AURA). **246-247 Moonpans.com:** (b). **248-249 Alamy Images:** Dennis Hallinan. **249 Dorling Kindersley:** NASA. **250 NASA:** SOHO / EIT Consortium / ESA. **251 NASA:** NASA and The Hubble Heritage Team (STScI / AURA) Acknowledgment: R. G. French (Wellesley College), J. Cuzzi (NASA / Ames), L. Dones (SwRI), and J. Lissauer (NASA / Ames) (br). **252-253 Corbis:** Bryan Allen. **253 Science Photo Library:** Chris Butler (tr). **254 Dorling Kindersley:** Bob Gathany (tr). **HubbleSite:** ESA, NASA, and Felix Mirabel (French Atomic Energy Commission and Institute for Astronomy and Space Physics / Conicet of Argentina) (bl). **255 Corbis:** Ed Darack/ Science Faction (br). **256 NASA:** JPL-Caltech / T. Pyle (SSC)

All other images © Dorling Kindersley

For further information see: www.dkimages.com